Introduction to Atmospheric Physics

Introduction to Atmospheric Physics

Editor: Francis Kenning

www.callistoreference.com

Callisto Reference,
118-35 Queens Blvd., Suite 400,
Forest Hills, NY 11375, USA

Visit us on the World Wide Web at:
www.callistoreference.com

ISBN: 978-1-64116-009-4 (Hardback)

Cataloging-in-Publication Data

Introduction to atmospheric physics / edited by Francis Kenning.
    p. cm.
Includes bibliographical references and index.
ISBN 978-1-64116-009-4
1. Atmospheric physics. 2. Geophysics. I. Kenning, Francis.
QC861.3 .I58 2018
551.5--dc23

Table of Contents

Preface

Atmospheric physics refers to the process of understanding and analyzing the atmosphere using the laws and elements of physics. It focuses on studying the atmosphere of other planets, by using techniques like radiation budget, fluid flow equations, cloud physics, spatial statistics, scattering theory, etc. It also plays a crucial role in deriving models to predict and understand varied weather patterns. This book aims to shed light on some of the unexplored aspects of atmospheric physics. Such selected concepts that redefine the subject have been presented in it. This textbook is an essential guide for both academicians and those who wish to pursue this discipline further.

To facilitate a deeper understanding of the contents of this book a short introduction of every chapter is written below:

Chapter 1- Meteorology studies the Earth's atmosphere and its weather conditions whereas oceanography studies the physical aspects of the ocean. The diverse topics related to the fields of meteorology and oceanography have been elaborately discussed in the section. This chapter will provide an integrated understanding of meteorology and oceanography.

Chapter 2- Weather is the state of the atmosphere of a place. Some of the common weather phenomena are currents, wind and rain. Weather can be predicted by using computer-based models which helps in considering all the facets related to the weather. This chapter will provide an integrated understanding of weather.

Chapter 3- The thermodynamic phenomena that are behind the formation of clouds, changing in weather and temperature are known as atmospheric thermodynamics. Evaporation and condensation are two phenomena that provide the transfer of energy. This field is important for the modeling of particular regions and calamities like hurricanes and tornadoes. The topics discussed in the chapter are of great importance to broaden the existing knowledge on atmospheric thermodynamics.

Chapter 4- Climate depends on the interaction of atmospheric processes with other Earth systems. Temperature and climatic conditions of the landmass is directly affected by the ocean. Atmospheric and oceanic physics is best understood in confluence with the major topics listed in the following chapter.

I owe the completion of this book to the never-ending support of my family, who supported me throughout the project.

Editor

Meteorology and Oceanography: An Integrated Study

Meteorology studies the Earth's atmosphere and its weather conditions whereas oceanography studies the physical aspects of the ocean. The diverse topics related to the fields of meteorology and oceanography have been elaborately discussed in the section. This chapter will provide an integrated understanding of meteorology and oceanography.

Weather and Climate

Sun is the fundamental source of energy, which heats the atmosphere and creates air motions. These air motions distribute heat and moisture in the atmosphere and weather systems are created whenever energy from sun combines with moisture. Hence, weather pertains to a state of the atmosphere that is experienced at a given time (also reported at synoptic hours) that is defined by variables such as temperature, pressure, winds, rainfall, cloud cover (expressed in octas) and several other dynamical and thermodynamical variables. But, climate is the mean state of the atmosphere that arises from interplay of meteorological processes that constantly change the state of the atmosphere in motion. Thus, climate would refer to averages of weather elements obtained from their time series for a location or any region. In effect, the climate could imply monthly, seasonal or annual mean distributions of temperature, rainfall or for that matter of any other weather element of interest, depending on the averaging period. In this sense weather and climate are intimately linked together. The day-to-day variations in weather if persist for longer periods, leave invariably their footprints in the climate of a place or region. Putatively, climate could also refer to the probability of occurrence of a particular type of weather regime during a particular period of the year to which the reference has been made. To illustrate this point, consider the monsoon rainfall in India. The country receives nearly 75% of its annual rainfall during the 120-day long monsoon season (June-September). The forthwith interpretation could be that the probability of rain water falling everyday during the monsoon season could be 75% of the daily All-India mean rainfall (temporal sense); or on the country scale, 75% area of the country has the probability of receiving average rainfall (spatial sense) daily in this 120-day period. Obviously, predictions based on persistence have high failure rate which brings into prominence the role of numerical weather prediction models. Nevertheless, the persistent structures in the ever-evolving patterns of atmospheric variables have to be studied for climate analysis; the variance of these variables from their weakly, monthly, seasonally or annually persistent structures precisely indicates the climate variability on the corresponding scales. The above description may however imply that the climate depends only on atmospheric processes. If accepted in totality, it would be misleading. Essentially, the interaction of the climate with other processes and components of the earth system gives rise to feedbacks that significantly influence the response of the climate system. For example, the variations in the

solar constant and the human-induced changes in the composition of the atmosphere would secu-
larly affect the radiative balance of the earth system and its climate; that is, the climatic response
would either amplify or dampen under the influence of any such imposed external forcing.

Since atmospheric models are backed by highly advanced data assimilation systems, major forecasting centres nowadays are issuing seasonal forecasts. With coupled ocean-atmosphere models, longer forecasts out to a decade are also possible and are being indeed produced to predict
phenomena arising from interaction of atmospheric and oceanic processes, such as El Niño and
Southern Oscillation. Such models require huge resources, computing power and analyses prepared from global atmospheric observations and ocean data of several years. Our understanding
of the climate system has grown steadily as a result of success in numerical simulations of weather
and climate as well as from careful process studies under three different fields, viz., meteorology,
oceanography and aeronomy, which practically developed independently of each other.

Meteorology

It is the science of atmosphere that deals with the changing patterns of heat, moisture and air motion in the three dimensional space, which can be described by physical laws. Understanding the
behaviour and distribution of heat and moisture is the key to understanding weather systems that
can also be probed by ground-based instruments and satellite observations. The pressure and wind
systems on the synoptic scale are analysed for weather prognosis; hence weather charts have played
a central role in synoptic forecasting of weather though it has limited range of validity. Sir Gilbert
Walker, however, invented statistical methods to make a long-range forecasting of monsoon rainfall.
Such forecasts for the All-India summer monsoon rainfall have been successful for the country as a
whole nearly for a century, though district-wise rainfall forecast with such techniques turns out to
be very difficult. Now, weather predictions are accurately made up to 10 days in advance by assim-
ilating conventional meteorological (mostly radiosonde) data and satellite radiances directly into
more sophisticated high-resolution three-dimensional dynamical models, which also serve as tools
of climate and climate change simulations. Interestingly, meteorology evolved with the theory that
successfully explained the advent of summer monsoons over India by Hadley (1857).

Meteorology is a branch of the atmospheric sciences which includes atmospheric chemistry and
atmospheric physics, with a major focus on weather forecasting. The study of meteorology dates
back millennia, though significant progress in meteorology did not occur until the 18th century. The 19th century saw modest progress in the field after weather observation networks were
formed across broad regions. Prior attempts at prediction of weather depended on historical data.
It wasn't until after the elucidation of the laws of physics and, more particularly, the development
of the computer, allowing for the automated solution of a great many equations that model the
weather, in the latter half of the 20th century that significant breakthroughs in weather forecasting
were achieved.

Meteorological phenomena are observable weather events that are explained by the science of meteorology. Meteorological phenomena are described and quantified by the variables of Earth's atmosphere: temperature, air pressure, water vapor, mass flow, and the variations and interactions

of those variables, and how they change over time. Different spatial scales are used to describe and predict weather on local, regional, and global levels.

Meteorology, climatology, atmospheric physics, and atmospheric chemistry are sub-disciplines of the atmospheric sciences. Meteorology and hydrology compose the interdisciplinary field of hydrometeorology. The interactions between Earth's atmosphere and its oceans are part of a coupled ocean-atmosphere system. Meteorology has application in many diverse fields such as the military, energy production, transport, agriculture, and construction.

The word "meteorology" is from Greek *metéōros* "lofty; high (in the sky)" and *-logia* "-(o)logy", i.e. "the study of things in the air".

History

Parhelion (sundog) in Savoie

The beginnings of meteorology can be traced back to ancient India, as the Upanishads contain serious discussion about the processes of cloud formation and rain and the seasonal cycles caused by the movement of Earth around the sun. Varāhamihira's classical work *Brihatsamhita*, written about 500 AD, provides clear evidence that a deep knowledge of atmospheric processes existed even in those times.

In 350 BC, Aristotle wrote *Meteorology*. Aristotle is considered the founder of meteorology. One of the most impressive achievements described in the *Meteorology* is the description of what is now known as the hydrologic cycle.

The book De Mundo (composed before 250 BC or between 350 and 200 BC) noted

> If the flashing body is set on fire and rushes violently to the Earth it is called a thunderbolt; if it is only half of fire, but violent also and massive, it is called a *meteor*; if it is entirely free from fire, it is called a smoking bolt. They are all called 'swooping bolts' because they swoop down upon the Earth. Lightning is sometimes smoky, and is then called 'smoldering lightning"; sometimes it darts quickly along, and is then said to be *vivid*. At other times, it travels in crooked lines, and is called *forked lightning*. When it swoops down upon some object it is called 'swooping lightning'.

The Greek scientist Theophrastus compiled a book on weather forecasting, called the *Book of Signs*. The work of Theophrastus remained a dominant influence in the study of weather and in weather forecasting for nearly 2,000 years. In 25 AD, Pomponius Mela, a geographer for the Roman Empire, formalized the climatic zone system. According to Toufic Fahd, around the 9th century, Al-Dinawari wrote the *Kitab al-Nabat* (*Book of Plants*), in which he deals with the application of meteorology to agriculture during the Muslim Agricultural Revolution. He describes the meteorological character of the sky, the planets and constellations, the sun and moon, the lunar phases indicating seasons and rain, the *anwa* (heavenly bodies of rain), and atmospheric phenomena such as winds, thunder, lightning, snow, floods, valleys, rivers, lakes.

Research of Visual Atmospheric Phenomena

Twilight at Baker Beach

Ptolemy wrote on the atmospheric refraction of light in the context of astronomical observations. In 1021, Alhazen showed that atmospheric refraction is also responsible for twilight; he estimated that twilight begins when the sun is 19 degrees below the horizon, and also used a geometric determination based on this to estimate the maximum possible height of the Earth's atmosphere as 52,000 *passim* (about 49 miles, or 79 km).

St. Albert the Great was the first to propose that each drop of falling rain had the form of a small sphere, and that this form meant that the rainbow was produced by light interacting with each raindrop. Roger Bacon was the first to calculate the angular size of the rainbow. He stated that a rainbow summit can not appear higher than 42 degrees above the horizon. In the late 13th century and early 14th century, Kamāl al-Dīn al-Fārisī and Theodoric of Freiberg were the first to give the correct explanations for the primary rainbow phenomenon. Theoderic went further and also explained the secondary rainbow. In 1716, Edmund Halley suggested that aurorae are caused by "magnetic effluvia" moving along the Earth's magnetic field lines.

Instruments and Classification Scales

In 1441, King Sejong's son, Prince Munjong, invented the first standardized rain gauge. These were sent throughout the Joseon Dynasty of Korea as an official tool to assess land taxes based upon a farmer's potential harvest. In 1450, Leone Battista Alberti developed a swinging-plate anemometer, and was known as the first *anemometer*. In 1607, Galileo Galilei constructed a thermoscope. In 1611, Johannes Kepler wrote the first scientific treatise on snow crystals: "Strena Seu de Nive Sexangula (A New Year's Gift of Hexagonal Snow)". In 1643, Evangelista Torricelli invented the mercury barometer. In 1662, Sir Christopher Wren invented the mechanical, self-emptying, tipping bucket rain gauge. In 1714, Gabriel Fahrenheit created a reliable scale for measuring temperature with a mercury-type thermometer. In 1742, Anders Celsius, a Swedish astronomer, proposed the "centigrade" temperature scale, the predecessor of the current Celsius scale. In 1783, the first hair hygrometer was demonstrated by Horace-Bénédict de Saussure. In 1802–1803, Luke Howard wrote *On the Modification of Clouds*, in which he assigns cloud types Latin names. In 1806, Francis Beaufort introduced his system for classifying wind speeds. Near the end of the 19th century the first cloud atlases were published, including the *International Cloud Atlas*, which has remained in print ever since. The April 1960 launch of the first successful weather satellite, TIROS-1, marked the beginning of the age where weather information became available globally.

THE ROBINSON ANEMOMETER.

A hemispherical cup anemometer

Atmospheric Composition Research

In 1648, Blaise Pascal rediscovered that atmospheric pressure decreases with height, and deduced that there is a vacuum above the atmosphere. In 1738, Daniel Bernoulli published *Hydrodynamics*, initiating the Kinetic theory of gases and established the basic laws for the theory of gases. In 1761, Joseph Black discovered that ice absorbs heat without changing its temperature when melting. In 1772, Black's student Daniel Rutherford discovered nitrogen, which he called *phlogisticated air*, and together they developed the phlogiston theory. In 1777, Antoine Lavoisier discovered oxygen and developed an explanation for combustion. In 1783, in Lavoisier's essay "Reflexions sur le phlogistique", he deprecates the phlogiston theory and proposes a caloric theory. In 1804, Sir John Leslie observed that a matte black surface radiates heat more effectively than a polished surface, suggesting the importance of black body radiation. In 1808, John Dalton defended caloric theory in *A New System of Chemistry* and described how it combines with matter, especially gases; he proposed that the heat capacity of gases varies inversely with atomic weight. In 1824, Sadi Carnot analyzed the efficiency of steam engines using caloric theory; he developed the notion of a reversible process and, in postulating that no such thing exists in nature, laid the foundation for the second law of thermodynamics.

Research into Cyclones and Air Flow

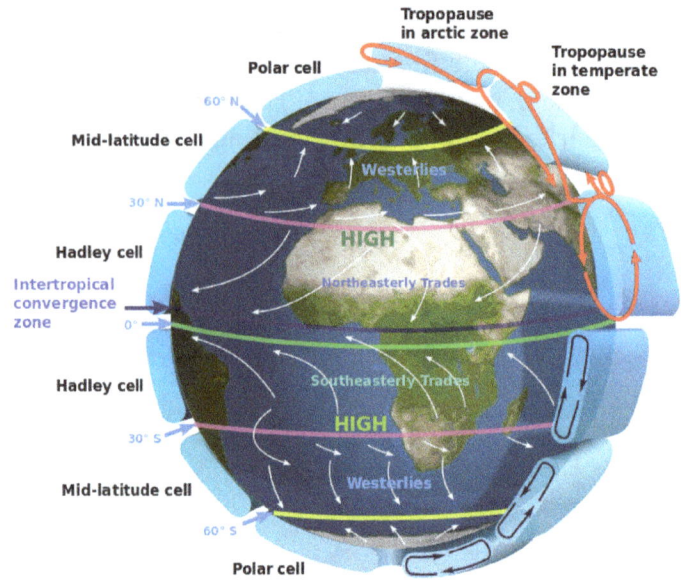

General Circulation of the Earth's Atmosphere: The westerlies and trade winds
are part of the Earth's atmospheric circulation

In 1494, Christopher Columbus experienced a tropical cyclone, which led to the first written European account of a hurricane. In 1686, Edmund Halley presented a systematic study of the trade winds and monsoons and identified solar heating as the cause of atmospheric motions. In 1735, an *ideal* explanation of global circulation through study of the trade winds was written by George Hadley. In 1743, when Benjamin Franklin was prevented from seeing a lunar eclipse by a hurricane, he decided that cyclones move in a contrary manner to the winds at their periphery. Understanding the kinematics of how exactly the rotation of the Earth affects airflow was partial at first. Gaspard-Gustave Coriolis published a paper in 1835 on the energy yield of machines with rotating parts, such as waterwheels. In 1856, William Ferrel proposed the existence of a circulation cell in the mid-latitudes, and the air within deflected by the Coriolis force resulting in the prevailing westerly winds. Late in the 19th century, the motion of air masses along isobars was understood to be the result of the large-scale interaction of the pressure gradient force and the deflecting force. By 1912, this deflecting force was named the Coriolis effect. Just after World War I, a group of meteorologists in Norway led by Vilhelm Bjerknes developed the Norwegian cyclone model that explains the generation, intensification and ultimate decay (the life cycle) of mid-latitude cyclones, and introduced the idea of fronts, that is, sharply defined boundaries between air masses. The group included Carl-Gustaf Rossby (who was the first to explain the large scale atmospheric flow in terms of fluid dynamics), Tor Bergeron (who first determined how rain forms) and Jacob Bjerknes.

Observation Networks and Weather Forecasting

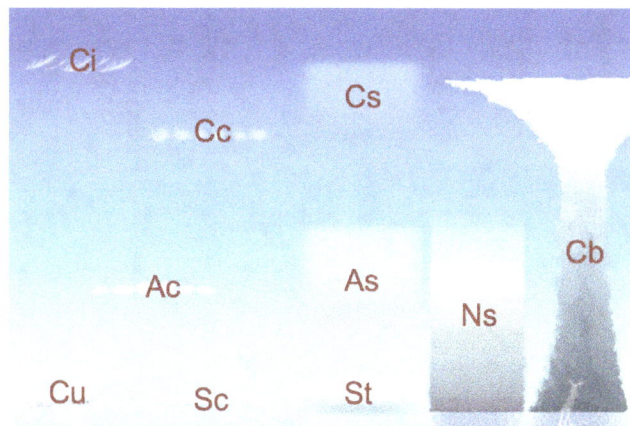

Cloud classification by altitude of occurrence

In 1654, Ferdinando II de Medici established the first *weather observing* network, that consisted of meteorological stations in Florence, Cutigliano, Vallombrosa, Bologna, Parma, Milan, Innsbruck, Osnabrück, Paris and Warsaw. The collected data were sent to Florence at regular time intervals. In 1832, an electromagnetic telegraph was created by Baron Schilling. The arrival of the electrical telegraph in 1837 afforded, for the first time, a practical method for quickly gathering surface weather observations from a wide area. This data could be used to produce maps of the state of the atmosphere for a region near the Earth's surface and to study how these states evolved through time. To make frequent weather forecasts based on these data required a reliable network of observations, but it was not until 1849 that the Smithsonian Institution began to establish an observation network across the United States under the leadership of Joseph Henry. Similar observation networks were established in Europe at this time. The Reverend William Clement Ley was key in understanding of cirrus clouds and early understandings of Jet Streams. Later after this Charles Kenneth Mackinnon Douglas known as 'CKM' Douglas read Ley's papers after his death and carried on the early study of weather systems. Nineteenth century researchers in meteorology were drawn from military or medical backgrounds, rather than trained as dedicated scientists. In 1854, the United Kingdom government appointed Robert FitzRoy to the new office of *Meteorological Statist to the Board of Trade* with the task of gathering weather observations at sea. FitzRoy's office became the United Kingdom Meteorological Office in 1854, the second oldest national meteorological service in the world (the Central Institution for Meteorology and Geodynamics ZAMG in Austria was founded in 1851 and is therefore the oldest weather service in the world). The first daily weather forecasts made by FitzRoy's Office were published in *The Times* newspaper in 1860. The following year a system was introduced of hoisting storm warning cones at principal ports when a gale was expected.

Over the next 50 years many countries established national meteorological services. The India Meteorological Department (1875) was established to follow tropical cyclone and monsoon. The Finnish Meteorological Central Office (1881) was formed from part of Magnetic Observatory of Helsinki University. Japan's Tokyo Meteorological Observatory, the forerunner of the Japan Meteorological Agency, began constructing surface weather maps in 1883. The United States Weather Bureau (1890) was established under the United States Department of Agriculture. The Australian Bureau of Meteorology (1906) was established by a Meteorology Act to unify existing state meteorological services.

Numerical Weather Prediction

A meteorologist at the console of the IBM 7090 in the Joint Numerical
Weather Prediction Unit. c. 1965

In 1904, Norwegian scientist Vilhelm Bjerknes first argued in his paper *Weather Forecasting as a Problem in Mechanics and Physics* that it should be possible to forecast weather from calculations based upon natural laws.

It was not until later in the 20th century that advances in the understanding of atmospheric physics led to the foundation of modern numerical weather prediction. In 1922, Lewis Fry Richardson published "Weather Prediction By Numerical Process", after finding notes and derivations he worked on as an ambulance driver in World War I. He described how small terms in the prognostic fluid dynamics equations that govern atmospheric flow could be neglected, and a numerical calculation scheme that could be devised to allow predictions. Richardson envisioned a large auditorium of thousands of people performing the calculations. However, the sheer number of calculations required was too large to complete without electronic computers, and the size of the grid and time steps used in the calculations led to unrealistic results. Though numerical analysis later found that this was due to numerical instability.

Starting in the 1950s, numerical forecasts with computers became feasible. The first weather forecasts derived this way used barotropic (single-vertical-level) models, and could successfully predict the large-scale movement of midlatitude Rossby waves, that is, the pattern of atmospheric lows and highs. In 1959, the UK Meteorological Office received its first computer, a Ferranti Mercury.

In the 1960s, the chaotic nature of the atmosphere was first observed and mathematically described by Edward Lorenz, founding the field of chaos theory. These advances have led to the current use of ensemble forecasting in most major forecasting centers, to take into account uncertainty arising from the chaotic nature of the atmosphere. Mathematical models used to predict the long term weather of the Earth (Climate models), have been developed that have a resolution today that are as coarse as the older weather prediction models. These climate models are used to investigate long-term climate shifts, such as what effects might be caused by human emission of greenhouse gases.

Meteorologists

Meteorologists are scientists who study meteorology. The American Meteorological Society published and continually updates an authoritative electronic *Meteorology Glossary*. Meteorologists work in government agencies, private consulting and research services, industrial enterprises, utilities, radio and television stations, and in education. In the United States, meteorologists held about 9,400 jobs in 2009.

Meteorologists are best known by the public for weather forecasting. Some radio and television weather forecasters are professional meteorologists, while others are reporters (weather specialist, weatherman, etc.) with no formal meteorological training. The American Meteorological Society and National Weather Association issue "Seals of Approval" to weather broadcasters who meet certain requirements.

Equipment

Satellite image of Hurricane Hugo with a polar low visible at the top of the image.

Each science has its own unique sets of laboratory equipment. In the atmosphere, there are many things or qualities of the atmosphere that can be measured. Rain, which can be observed, or seen anywhere and anytime was one of the first atmospheric qualities measured historically. Also, two other accurately measured qualities are wind and humidity. Neither of these can be seen but can be felt. The devices to measure these three sprang up in the mid-15th century and were respectively the rain gauge, the anemometer, and the hygrometer. Many attempts had been made prior to the 15th century to construct adequate equipment to measure the many atmospheric variables. Many were faulty in some way or were simply not reliable. Even Aristotle noted this in some of his work as the difficulty to measure the air.

Sets of surface measurements are important data to meteorologists. They give a snapshot of a variety of weather conditions at one single location and are usually at a weather station, a ship or a weather buoy. The measurements taken at a weather station can include any number of atmospheric observables. Usually, temperature, pressure, wind measurements, and humidity are the variables that are measured by a thermometer, barometer, anemometer, and hygrometer, respectively. Professional stations may also include air quality sensors (carbon monoxide, carbon dioxide, methane, ozone, dust, and smoke), ceilometer (cloud ceiling), falling precipitation sensor, flood sensor, lightning sensor, microphone (explosions, sonic booms, thunder), pyranometer/pyrheliometer/spectroradiometer (IR/Vis/UV photodiodes), rain gauge/snow gauge, scintillation counter (background radiation, fallout, radon), seismometer (earthquakes and tremors), transmissometer (visibility), and a GPS clock for data logging. Upper air data are of crucial importance for weather forecasting. The most widely used technique is launches of radiosondes. Supplementing the radiosondes a network of aircraft collection is organized by the World Meteorological Organization.

Remote sensing, as used in meteorology, is the concept of collecting data from remote weather events and subsequently producing weather information. The common types of remote sensing are Radar, Lidar, and satellites (or photogrammetry). Each collects data about the atmosphere from a remote location and, usually, stores the data where the instrument is located. Radar and Lidar are not passive because both use EM radiation to illuminate a specific portion of the atmosphere. Weather satellites along with more general-purpose Earth-observing satellites circling the earth at various altitudes have become an indispensable tool for studying a wide range of phenomena from forest fires to El Niño.

Spatial Scales

The study of the atmosphere can be divided into distinct areas that depend on both time and spatial scales. At one extreme of this scale is climatology. In the timescales of hours to days, meteorology separates into micro-, meso-, and synoptic scale meteorology. Respectively, the geospatial size of each of these three scales relates directly with the appropriate timescale.

Other subclassifications are used to describe the unique, local, or broad effects within those subclasses.

| Typical Scales of Atmospheric Motion Systems ||
Type of motion	Horizontal scale (meter)
Molecular mean free path	10^{-3}
Minute turbulent eddies	10^{-2} - 10^{-1}
Small eddies	10^{-1} - 1
Dust devils	1 - 10
Gusts	10 - 10^2
Tornadoes	10^2

Thunderclouds	10^3
Fronts, squall lines	$10^4 - 10^5$
Hurricanes	10^5
Synoptic Cyclones	10^6
Planetary waves	10^7
Atmospheric tides	10^7
Mean zonal wind	10^7

Microscale

Microscale meteorology is the study of atmospheric phenomena on a scale of about 1 kilometre (0.62 mi) or less. Individual thunderstorms, clouds, and local turbulence caused by buildings and other obstacles (such as individual hills) are modeled on this scale.

Mesoscale

Mesoscale meteorology is the study of atmospheric phenomena that has horizontal scales ranging from 1 km to 1000 km and a vertical scale that starts at the Earth's surface and includes the atmospheric boundary layer, troposphere, tropopause, and the lower section of the stratosphere. Mesoscale timescales last from less than a day to weeks. The events typically of interest are thunderstorms, squall lines, fronts, precipitation bands in tropical and extratropical cyclones, and topographically generated weather systems such as mountain waves and sea and land breezes.

Synoptic Scale

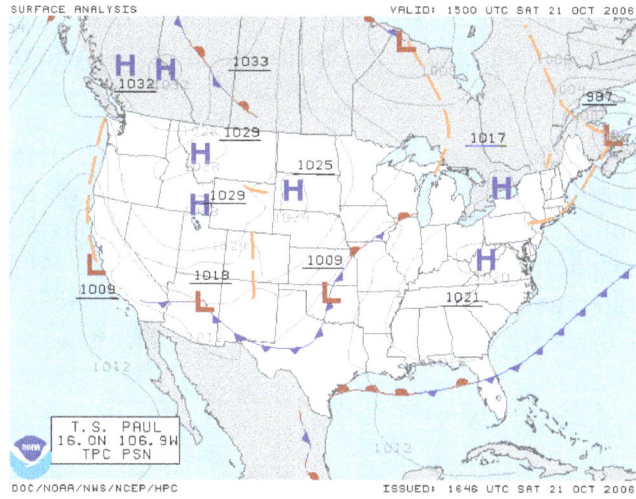

NOAA: Synoptic scale weather analysis.

Synoptic scale meteorology predicts atmosperic changes at scales up to 1000 km and 10^5 sec (28 days), in time and space. At the synoptic scale, the Coriolis acceleration acting on moving air masses (outside of the tropics) plays a dominant role in predictions. The phenomena typically described by synoptic meteorology include events such as extratropical cyclones, baroclinic troughs and ridges, frontal zones, and to some extent jet streams. All of these are typically given on weather maps for a specific time. The minimum horizontal scale of synoptic phenomena is limited to the spacing between surface observation stations.

Global Scale

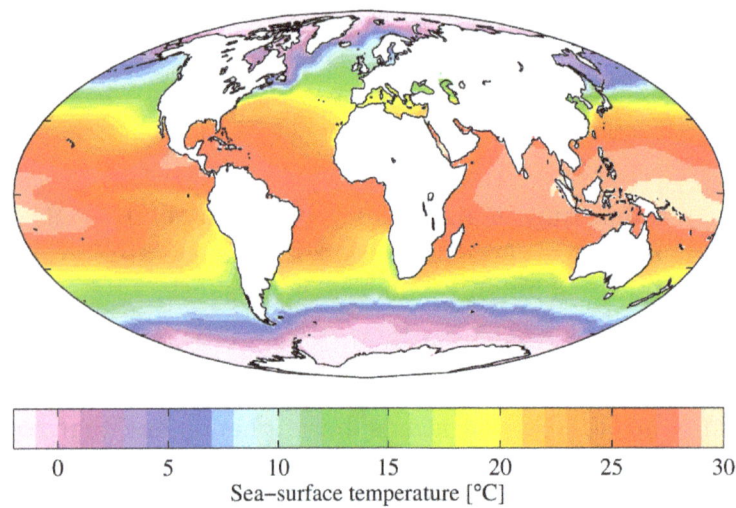

Annual mean sea surface temperatures.

Global scale meteorology is the study of weather patterns related to the transport of heat from the tropics to the poles. Very large scale oscillations are of importance at this scale. These oscillations have time periods typically on the order of months, such as the Madden–Julian oscillation, or years, such as the El Niño–Southern Oscillation and the Pacific decadal oscillation. Global scale meteorology pushes into the range of climatology. The traditional definition of climate is pushed into larger timescales and with the understanding of the longer time scale global oscillations, their effect on climate and weather disturbances can be included in the synoptic and mesoscale times-cales predictions.

Numerical Weather Prediction is a main focus in understanding air–sea interaction, tropical meteorology, atmospheric predictability, and tropospheric/stratospheric processes. The Naval Research Laboratory in Monterey, California, developed a global atmospheric model called Navy Operational Global Atmospheric Prediction System (NOGAPS). NOGAPS is run operationally at Fleet Numerical Meteorology and Oceanography Center for the United States Military. Many other global atmospheric models are run by national meteorological agencies.

Some Meteorological Principles

Boundary Layer Meteorology

Boundary layer meteorology is the study of processes in the air layer directly above Earth's surface, known as the atmospheric boundary layer (ABL). The effects of the surface – heating, cooling, and friction – cause turbulent mixing within the air layer. Significant movement of heat, matter, or momentum on time scales of less than a day are caused by turbulent motions. Boundary layer meteorology includes the study of all types of surface–atmosphere boundary, including ocean, lake, urban land and non-urban land for the study of meteorology.

Dynamic Meteorology

Dynamic meteorology generally focuses on the fluid dynamics of the atmosphere. The idea of air parcel is used to define the smallest element of the atmosphere, while ignoring the discrete molecular and chemical nature of the atmosphere. An air parcel is defined as a point in the fluid continuum of the atmosphere. The fundamental laws of fluid dynamics, thermodynamics, and motion are used to study the atmosphere. The physical quantities that characterize the state of the atmosphere are temperature, density, pressure, etc. These variables have unique values in the continuum.

Applications

Weather Forecasting

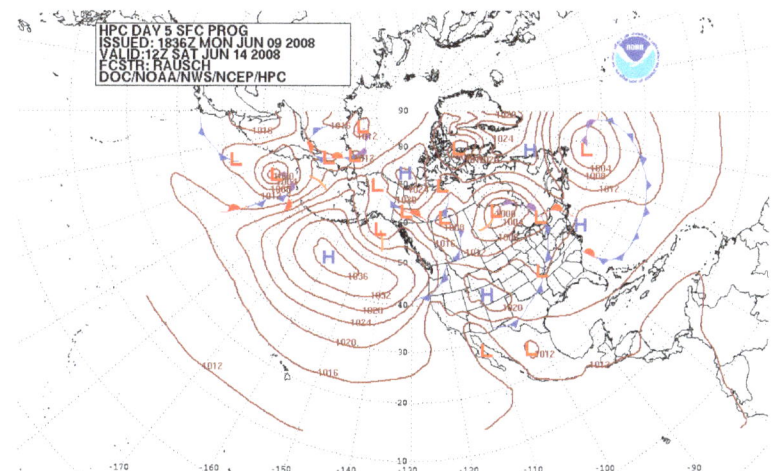

Forecast of surface pressures five days into the future for the north Pacific, North America, and north Atlantic Ocean

Weather forecasting is the application of science and technology to predict the state of the atmosphere at a future time and given location. Humans have attempted to predict the weather informally for millennia and formally since at least the 19th century. Weather forecasts are made by collecting quantitative data about the current state of the atmosphere and using scientific understanding of atmospheric processes to project how the atmosphere will evolve.

Once an all-human endeavor based mainly upon changes in barometric pressure, current weather conditions, and sky condition, forecast models are now used to determine future conditions. Human input is still required to pick the best possible forecast model to base the forecast upon, which involves pattern recognition skills, teleconnections, knowledge of model performance, and knowledge of model biases. The chaotic nature of the atmosphere, the massive computational power required to solve the equations that describe the atmosphere, error involved in measuring the initial conditions, and an incomplete understanding of atmospheric processes mean that forecasts become less accurate as the difference in current time and the time for which the forecast is being made (the *range* of the forecast) increases. The use of ensembles and model consensus help narrow the error and pick the most likely outcome.

There are a variety of end uses to weather forecasts. Weather warnings are important forecasts because they are used to protect life and property. Forecasts based on temperature and precipitation are important to agriculture, and therefore to commodity traders within stock markets. Temperature forecasts are used by utility companies to estimate demand over coming days. On an everyday basis, people use weather forecasts to determine what to wear. Since outdoor activities are severely curtailed by heavy rain, snow, and wind chill, forecasts can be used to plan activities around these events, and to plan ahead and survive them.

Aviation Meteorology

Aviation meteorology deals with the impact of weather on air traffic management. It is important for air crews to understand the implications of weather on their flight plan as well as their aircraft, as noted by the Aeronautical Information Manual:

The effects of ice on aircraft are cumulative—thrust is reduced, drag increases, lift lessens, and weight increases. The results are an increase in stall speed and a deterioration of aircraft performance. In extreme cases, 2 to 3 inches of ice can form on the leading edge of the airfoil in less than 5 minutes. It takes but 1/2 inch of ice to reduce the lifting power of some aircraft by 50 percent and increases the frictional drag by an equal percentage.

Agricultural Meteorology

Meteorologists, soil scientists, agricultural hydrologists, and agronomists are persons concerned with studying the effects of weather and climate on plant distribution, crop yield, water-use efficiency, phenology of plant and animal development, and the energy balance of managed and natural ecosystems. Conversely, they are interested in the role of vegetation on climate and weather.

Hydrometeorology

Hydrometeorology is the branch of meteorology that deals with the hydrologic cycle, the water budget, and the rainfall statistics of storms. A hydrometeorologist prepares and issues forecasts of accumulating (quantitative) precipitation, heavy rain, heavy snow, and highlights areas with the potential for flash flooding. Typically the range of knowledge that is required overlaps with climatology, mesoscale and synoptic meteorology, and other geosciences.

The multidisciplinary nature of the branch can result in technical challenges, since tools and solutions from each of the individual disciplines involved may behave slightly differently, be optimized for different hard- and software platforms and use different data formats. There are some initiatives - such as the DRIHM project - that are trying to address this issue.

Nuclear Meteorology

Nuclear meteorology investigates the distribution of radioactive aerosols and gases in the atmosphere.

Maritime Meteorology

Maritime meteorology deals with air and wave forecasts for ships operating at sea. Organizations such as the Ocean Prediction Center, Honolulu National Weather Service forecast office, United Kingdom Met Office, and JMA prepare high seas forecasts for the world's oceans.

Military Meteorology

Military meteorology is the research and application of meteorology for military purposes. In the United States, the United States Navy's Commander, Naval Meteorology and Oceanography Command oversees meteorological efforts for the Navy and Marine Corps while the United States Air Force's Air Force Weather Agency is responsible for the Air Force and Army.

Environmental Meteorology

Environmental meteorology mainly analyzes industrial pollution dispersion physically and chemically based on meteorological parameters such as temperature, humidity, wind, and various weather conditions.

Renewable Energy

Meteorology applications in renewable energy includes basic research, "exploration", and potential mapping of wind power and solar radiation for wind and solar energy.

Oceanography

It is the study of The Real Ocean often described as complex, dilute solution of extremely large volume of waters where several chemical reactions are taking place. In response to energy received from the atmosphere or through the atmosphere (i.e. solar radiation), circulations at various temporal and spatial scales characterize the ocean state. The ocean currents transport waters with dissolved material from one place to another. The life forms in the ocean depend upon the chemistry of their environment. Their distribution and productivity are determined by circulation and physical properties of ocean waters. However, for a complete understanding, the biology, chemistry, geology, and physics of the oceans must be known. Like atmosphere, ocean observations too are key to understanding oceanic phenomena and ocean currents. The ocean observations are compiled and preserved as hydrographical data that provide information on water depths, shorelines, tides, currents, bottom types, undersea ridges, valleys and other features in the realm of ocean.

Oceanography also known as oceanology, is the study of the physical and the biological aspects of the ocean. It is an Earth science covering a wide range of topics, including ecosystem dynamics; ocean currents, waves, and geophysical fluid dynamics; plate tectonics and the geology of the sea floor; and fluxes of various chemical substances and physical properties within the ocean and across its boundaries. These diverse topics reflect multiple disciplines that oceanographers blend to further knowledge of the world ocean and understanding of processes within: astronomy, biology, chemistry, climatology, geography, geology, hydrology, meteorology and physics. Paleoceanography studies the history of the oceans in the geologic past.

History

Map of the Gulf Stream by Benjamin Franklin, 1769-1770. Courtesy of the NOAA Photo Library.

Early History

Humans first acquired knowledge of the waves and currents of the seas and oceans in pre-historic times. Observations on tides were recorded by Aristotle and Strabo. Early exploration of the oceans was primarily for cartography and mainly limited to its surfaces and of the animals that fishermen brought up in nets, though depth soundings by lead line were taken.

Although Juan Ponce de León in 1513 first identified the Gulf Stream, and the current was well-known to mariners, Benjamin Franklin made the first scientific study of it and gave it its name. Franklin measured water temperatures during several Atlantic crossings and correctly explained the Gulf Stream's cause. Franklin and Timothy Folger printed the first map of the Gulf Stream in 1769-1770.

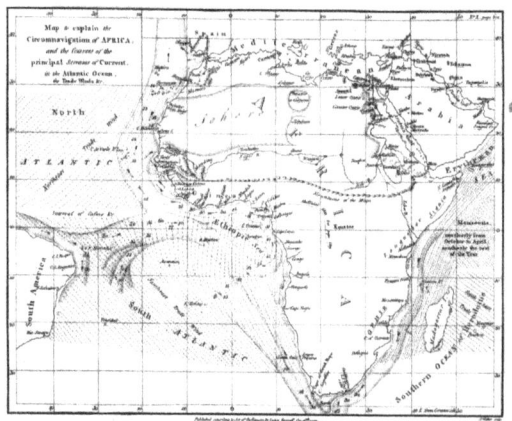

1799 map of the currents in the Atlantic and Indian Oceans, by James Rennell

Information on the currents of the Pacific Ocean was gathered by explorers of the late 18th century, including James Cook and Louis Antoine de Bougainville. James Rennell wrote the first scientific textbooks on oceanography, detailing the current flows of the Atlantic and Indian oceans. During a voyage around the Cape of Good Hope in 1777, he mapped *"the banks and currents at the Lagullas"*. He was also the first to understand the nature of the intermittent current near the Isles of Scilly, (now known as Rennell's Current).

Sir James Clark Ross took the first modern sounding in deep sea in 1840, and Charles Darwin published a paper on reefs and the formation of atolls as a result of the Second voyage of HMS Beagle in 1831-6. Robert FitzRoy published a four-volume report of the Beagle's three voyages. In 1841–1842 Edward Forbes undertook dredging in the Aegean Sea that founded marine ecology.

The first superintendent of the United States Naval Observatory (1842–1861), Matthew Fontaine Maury devoted his time to the study of marine meteorology, navigation, and charting prevailing winds and currents. His 1855 textbook *Physical Geography of the Sea* was one of the first comprehensive oceanography studies. Many nations sent oceanographic observations to Maury at the Naval Observatory, where he and his colleagues evaluated the information and distributed the results worldwide.

Modern Oceanography

Despite all this, human knowledge of the oceans remained confined to the topmost few fathoms of the water and a small amount of the bottom, mainly in shallow areas. Almost nothing was known of the ocean depths. The Royal Navy's efforts to chart all of the world's coastlines in the mid-19th century reinforced the vague idea that most of the ocean was very deep, although little more was known. As exploration ignited both popular and scientific interest in the polar regions and Africa, so too did the mysteries of the unexplored oceans.

H.M.S. CHALLENGER UNDER SAIL, 1874.

HMS *Challenger* undertook the first global marine research expedition in 1872.

The seminal event in the founding of the modern science of oceanography was the 1872-76 Challenger expedition. As the first true oceanographic cruise, this expedition laid the groundwork for an entire academic and research discipline. In response to a recommendation from the Royal Society, The British Government announced in 1871 an expedition to explore world's oceans and conduct appropriate scientific investigation. Charles Wyville Thompson and Sir John Murray launched the Challenger expedition. The Challenger, leased from the Royal Navy, was modified for scientific work and equipped with separate laboratories for natural history and chemistry. Under the scientific supervision of Thomson, Challenger travelled nearly 70,000 nautical miles (130,000 km) surveying and exploring. On her journey circumnavigating the globe, 492 deep sea soundings, 133 bottom dredges, 151 open water trawls and 263 serial water temperature observations were taken. Around 4,700 new species of marine life were discovered. The result was the *Report Of The Scientific Results of the Exploring Voyage of H.M.S. Challenger during the years 1873-76*. Murray, who supervised the publication, described the report as "the greatest advance in the knowledge of our planet since the celebrated discoveries of the fifteenth and sixteenth centuries". He went on to found the academic discipline of oceanography at the University of Edinburgh, which remained the centre for oceanographic research well into the 20th century. Murray was the first to study marine trenches and in particular the Mid-Atlantic Ridge, and map the sedimentary deposits in the oceans. He tried to map out the world's ocean currents based on salinity and temperature observations, and was the first to correctly understand the nature of coral reef development.

In the late 19th century, other Western nations also sent out scientific expeditions (as did private individuals and institutions). The first purpose built oceanographic ship, the Albatros, was built in 1882. In 1893, Fridtjof Nansen allowed his ship, Fram, to be frozen in the Arctic ice. This enabled him to obtain oceanographic, meteorological and astronomical data at a stationary spot over an extended period.

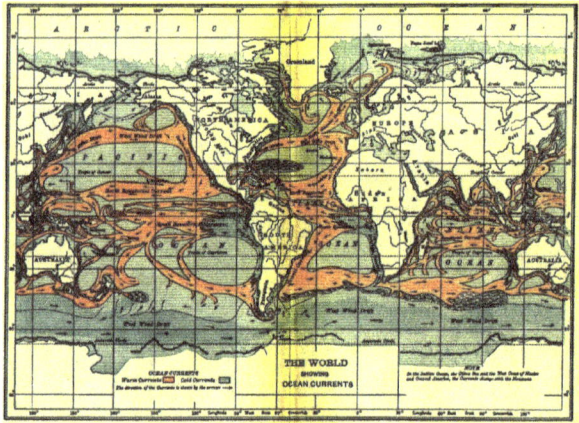

Ocean currents (1911)

Between 1907 and 1911 Otto Krümmel published the *Handbuch der Ozeanographie*, which became influential in awakening public interest in oceanography. The four-month 1910 North Atlantic expedition headed by John Murray and Johan Hjort was the most ambitious research oceanographic and marine zoological project ever mounted until then, and led to the classic 1912 book *The Depths of the Ocean*.

The first acoustic measurement of sea depth was made in 1914. Between 1925 and 1927 the "Meteor" expedition gathered 70,000 ocean depth measurements using an echo sounder, surveying the Mid-Atlantic ridge.

Sverdrup, Johnson and Fleming published *The Oceans* in 1942, which was a major landmark. *The Sea* (in three volumes, covering physical oceanography, seawater and geology) edited by M.N. Hill was published in 1962, while Rhodes Fairbridge's *Encyclopedia of Oceanography* was published in 1966.

The Great Global Rift, running along the Mid Atlantic Ridge, was discovered by Maurice Ewing and Bruce Heezen in 1953; in 1954 a mountain range under the Arctic Ocean was found by the Arctic Institute of the USSR. The theory of seafloor spreading was developed in 1960 by Harry Hammond Hess. The Ocean Drilling Program started in 1966. Deep sea vents were discovered in 1977 by John Corlis and Robert Ballard in the submersible DSV *Alvin*.

In the 1950s, Auguste Piccard invented the bathyscaphe and used the *Trieste* to investigate the ocean's depths. The United States nuclear submarine Nautilus made the first journey under the ice to the North Pole in 1958. In 1962 the FLIP (Floating Instrument Platform), a 355-foot spar buoy, was first deployed.

From the 1970s, there has been much emphasis on the application of large scale computers to oceanography to allow numerical predictions of ocean conditions and as a part of overall environmental change prediction. An oceanographic buoy array was established in the Pacific to allow prediction of El Niño events.

1990 saw the start of the World Ocean Circulation Experiment (WOCE) which continued until 2002. Geosat seafloor mapping data became available in 1995.

In recent years studies advanced particular knowledge on ocean acidification, ocean heat content, ocean currents, the El Niño phenomenon, mapping of methane hydrate deposits, the carbon cycle, coastal erosion, weathering and climate feedbacks in regards to climate change interactions.

Study of the oceans is linked to understanding global climate changes, potential global warming and related biosphere concerns. The atmosphere and ocean are linked because of evaporation and precipitation as well as thermal flux (and solar insolation). Wind stress is a major driver of ocean currents while the ocean is a sink for atmospheric carbon dioxide. All these factors relate to the ocean's biogeochemical setup.

Branches

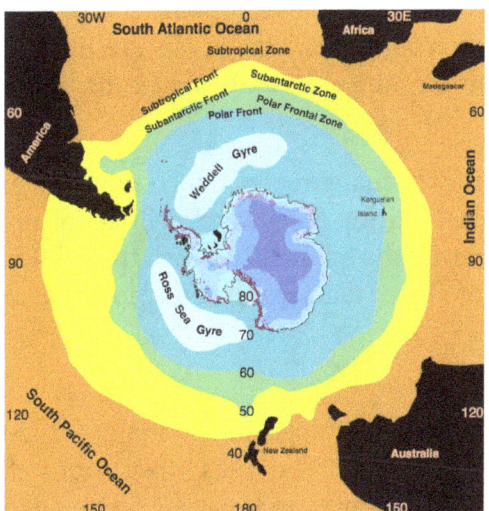

Oceanographic frontal systems on the Southern Hemisphere

The study of oceanography is divided into these four branches:

- Biological oceanography, or marine biology, investigates the ecology of marine organisms in the context of the physical, chemical, and geological characteristics of their ocean environment and the biology of individual marine organisms.

- Chemical oceanography and ocean chemistry, are the study of the chemistry of the ocean. Whereas chemical oceanography is primarily occupied with the study and understanding of seawater properties and its changes, focuses ocean chemistry primarily on the geochemical cycles.

- Geological oceanography, or marine geology, is the study of the geology of the ocean floor including plate tectonics and paleoceanography.

- Physical oceanography, or marine physics, studies the ocean's physical attributes including temperature-salinity structure, mixing, surface waves, internal waves, surface tides, internal tides, and currents.

Ocean Acidification

Ocean acidification describes the decrease in ocean pH that is caused by anthropogenic carbon dioxide (CO_2) emissions into the atmosphere. Seawater is slightly alkaline and had a preindustrial pH of about 8.2. More recently, anthropogenic activities have steadily increased the carbon dioxide content of the atmosphere; about 30–40% of the added CO_2 is absorbed by the oceans, forming carbonic acid and lowering the pH (now below 8.1) through ocean acidification. The pH is expected to reach 7.7 by the year 2100.

An important element for the skeletons of marine animals is calcium, but calcium carbonate becomes more soluble with pressure, so carbonate shells and skeletons dissolve below the carbonate compensation depth. Calcium carbonate becomes more soluble at lower pH, so ocean acidification is likely to affect marine organisms with calcareous shells, such as oysters, clams, sea urchins and corals, and the carbonate compensation depth will rise closer to the sea surface. Affected planktonic organisms will include pteropods, coccolithophorids and foraminifera, all important in the food chain. In tropical regions, corals are likely to be severely affected as they become less able to build their calcium carbonate skeletons, in turn adversely impacting other reef dwellers.

The current rate of ocean chemistry change seems to be unprecedented in Earth's geological history, making it unclear how well marine ecosystems will adapt to the shifting conditions of the near future. Of particular concern is the manner in which the combination of acidification with the expected additional stressors of higher temperatures and lower oxygen levels will impact the seas.

Ocean Currents

Since the early ocean expeditions in oceanography, a major interest was the study of the ocean currents and temperature measurements. The tides, the Coriolis effect, changes in direction and strength of wind, salinity and temperature are the main factors determining ocean currents. The thermohaline circulation (THC) *thermo-* referring to temperature and *-haline* referring to salt content connects 4 of 5 ocean basins and is primarily dependent on the density of sea water. Ocean currents such as the Gulf Stream are wind-driven surface currents.

Ocean Heat Content

Oceanic heat content (OHC) refers to the heat stored in the ocean. The changes in the ocean heat play an important role in sea level rise, because of thermal expansion. Ocean warming accounts for 90% of the energy accumulation from global warming between 1971 and 2010.

Oceanographic Institutions

Oceanographic Museum

The first international organization of oceanography was created in 1902 as the International Council for the Exploration of the Sea. In 1903 the Scripps Institution of Oceanography was founded, followed by Woods Hole Oceanographic Institution in 1930, Virginia Institute of Marine Science in 1938, and later the Lamont-Doherty Earth Observatory at Columbia University, and the School of Oceanography at University of Washington. In Britain, the National Oceanography Centre (an institute of the Natural Environment Research Council) is the successor to the UK's Institute of Oceanographic Sciences. In Australia, CSIRO Marine and Atmospheric Research (CMAR), is a leading centre. In 1921 the International Hydrographic Bureau (IHB) was formed in Monaco.

Aeronomy

As a subject, it deals with the study of atmospheres of planets; therefore, it encompasses topics like air chemistry, aerosols, clouds and radiation, and troposphere-stratospheric interactions. The middle atmosphere extending from 10-16 km to 100 km has been thoroughly investigated by aeronomists. The ozone problem has been at the heart of this subject, which led to the stratospheric ozone recovery strategies by successfully identifying the chemicals that are responsible for the destruction of ozone. Several nations united to protect the ozone layer once the ozone hole was discovered over Antarctica, which is as large as the size of North America during a given astral spring. The global warming induced by greenhouse gases (GHG) is currently the key topic of research and development of GHG emission control technologies. In the present day numerical prediction models, ozone is assimilated in the analysis like any other conventional weather parameter. The transboundary pollutant transport is one of the key concerns of today as NOx, SOx and other pollutants once emitted in the atmosphere disperse globally and trigger complex chemical reactions under the prevailing atmospheric conditions to form new particles which on ageing become cloud condensation nuclei that impact the growth and reflectivity of clouds, and precipitation intensities. It is worth pointing out that aerosols produce both direct (scattering of incoming solar radiation) and indirect forcing (through cloud albedo enhancement and inhibition of precipitation from clouds) on climate. The chemical reactions in the lowest layers (troposphere and stratosphere) are the focus of climate change research. The middle atmosphere dynamics that successfully explained the sudden stratospheric warming and quasi-biennial oscillations (QBO) in the stratosphere is at the core of aeronomy.

Aeronomy is the meteorological science of the upper region of the Earth's or other planetary atmospheres, which relates to the atmospheric motions, its chemical composition and properties, and the reaction to it from the environment from space. The term *aeronomy* was introduced by Sydney Chapman in a Letter to the Editor of *Nature* entitled *Some Thoughts on Nomenclature* in 1946. Studies within the subject also investigates the causes of dissociation or ionization processes.

Today the term also includes the science of the corresponding regions of the atmospheres of other planets. Aeronomy is a branch of atmospheric physics. Research in aeronomy requires access to balloons, satellites, and sounding rockets which provide valuable data about this region of the atmosphere. Atmospheric tides dominate the dynamics of the mesosphere and lower thermosphere, essential to understanding the atmosphere as a whole. Other phenomena studied are upper-atmospheric lightning discharges, such as red sprites, sprite halos or blue jets.

Atmospheric Tides

Atmospheric tides are global-scale periodic oscillations of the atmosphere. In many ways they are analogous to ocean tides. Atmospheric tides form an important mechanism for transporting energy input into the lower atmosphere from the upper atmosphere, while dominating the dynamics of the mesosphere and lower thermosphere. Therefore, learning about atmospheric tides is essential in understanding the atmosphere as a whole. Modeling and observations of atmospheric tides are needed in order to monitor and predict changes in the Earth's atmosphere.

Upper-atmospheric Lightning

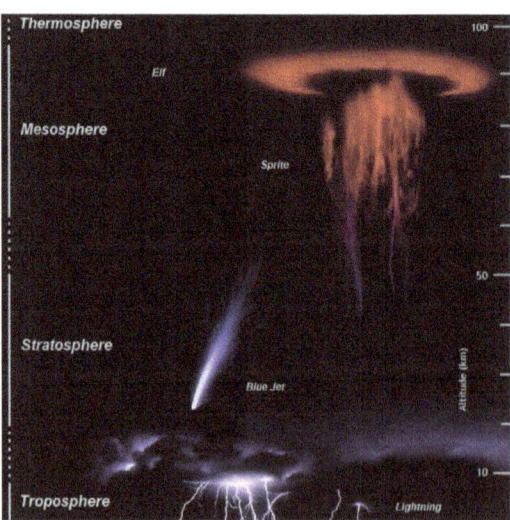

Representation of upper-atmospheric lightning and electrical-discharge phenomena

Upper-atmospheric lightning or upper-atmospheric discharge are terms sometimes used by researchers to refer to a family of electrical-breakdown phenomena that occur well above the altitudes of normal lightning. The preferred current usage is transient luminous events (TLEs) to refer to the various types of electrical-discharge phenomena induced in the upper atmosphere by tropospheric lightning. TLEs include red sprites, sprite halos, blue jets, and elves.

Atmosphere and Ocean in Perspective

What distinguishes the living planet from others in the solar system is the presence of water on earth, both in vapour form in the atmosphere and in liquid form in oceans. It also exists in the solid form (ice or snow) in the earth system. Whereas the atmosphere around the planet earth occupies a tiny fraction of the boundless space, the oceans occupy nearly 72% of the surface of the earth. Variation of density, temperature and other state variables (which we discuss as the course progresses) depict the three dimensional structure of the atmosphere and ocean primarily inferred from atmospheric and oceanographic observations. However, this three-dimensional structure is now being studied by two-dimensional satellite imageries. Before the advent of satellites, explorers have unfolded the mysteries of the space (atmosphere included) and of the oceans with their sheer grit and determination. The atmosphere that plays an important role in the day-to-day weather is relatively a shallow envelope of air surrounding the earth. Likewise, the oceans when compared with features of the earth surface are very deep, but are extremely shallow when they are related to their horizontal extent. Thus, both ocean and atmosphere are fluids in a state of turbulent motion and may be regarded as two-dimensional shallow layers. This shallow fluid layer is constantly in motion and the equations describing these motions are referred to as "shallow water equations". These equations have played a very significant role in formulating the scientific foundations of atmospheric and ocean dynamics. There is however a key difference in the height of the shallow layers: atmospheric shallow layer is regarded as a layer of average height of 10 km while the oceanic shallow layer has an average depth of 3.6 km. Similarly densities also differ by orders: density of air ~1.225 kg m^{-3} as compared to that of water ~1000 kg m^{-3}. Thus the mass of the oceans is ~250 times as large as that of the atmosphere. Also, atmosphere is transparent to electromagnetic radiation in certain wavelengths (window regions), but ocean is opaque to all forms of electromagnetic radiation.

For understanding weather and climate of the earth, observations on weather elements are of paramount importance. Temperature and wind are daily measured at regular intervals. The balloon flights carrying radiosondes aloft provide accurate description of thermodynamic variables like temperature, pressure and moisture. Satellite observations provide finer structure and high-resolution spatial (one kilometre) and temporal (half-hourly) details of temperature and moisture fields horizontally and, in general, of the atmospheric circulation. Ocean temperatures are measured by travelling ships and by sensors deployed in ocean waters and on satellites. While radiosondes provide records of atmospheric temperature, humidity and pressure distribution in the vertical, drifting floats make continuous measurements of salinity (conductivity), temperature and pressure (depth) in the ocean. These floats, also called the "floating CTDs", make same measurements that a Conductivity-Temperature-Depth (CTD) sensor does. They have the ability to sink typically to a depth of 2,000 metres in the ocean by changing their buoyancy and remain 7-10 days at that depth while continuously drifting with the ocean currents. By again changing their buoyancy, the drifting floats come to the surface and transmit the time of observations, position in the ocean and profiles of ocean variables to the orbiting satellites. These floats are indeed the marvels of technology and form the backbone of the Argo Programme in the global ocean. There are more than 3000 Argo floats that are active globally; roughly 10% of them are deployed in the Indian Ocean. In this course, we shall deal with physical oceanography that concerns to physical properties and processes within the ocean depths, and some conceptual models of ocean dynamics in order to explain surface currents and their relation to buoyancy and mixing.

Thermal Structure and Distribution of Atmospheric Constituents

The planet Earth is unique in the Solar System for sustaining life due to the presence of water and oxygen with several other gaseous constituents prominent among them being nitrogen, carbon dioxide and ozone in its atmosphere. Water exists in all three forms (solid, liquid and gas) on earth and transformation of water vapour into liquid form as cloud drops in the atmosphere, produces life giving rains falling back onto surface. Ozone absorbs the ultra-violet radiation from Sun and thus protects the lives of humans and animals on this planet from potential damages. The atmospheric pressure decreases with height but temperature shows strong variations in the vertical, typically increasing or decreasing linearly in a particular atmospheric layer. The temperature decreases linearly with height in the lowest atmospheric layer (called the troposphere) from surface to 10 km altitude and then increases in the "stratosphere" up to an altitude of 50 km.

The gradient of temperature in the vertical (or lapse rate) changes its sign from negative to positive in the "tropopause" region at about 10 km (100 hPa) above the ground, whereas the opposite happens (i.e. from positive to negative) in the "stratopause" region located at about 50 km (1 hPa) above the ground. Thus 99.0% of the mass of the atmosphere resides in the gaseous envelope extending from surface up to 50 km height that surrounds the planet earth. The region above 50 km is known as the "mesosphere" which is topped by the "mesopause", and the atmospheric pressure has reduced there to 0.001 hPa. The middle atmosphere refers invariably to both stratosphere and mesosphere that extends from tropopause to a height of 85 km.

Like in altitude, atmospheric temperatures also show strong variations from equator to pole: warmest at equator and coldest at the poles. The zonally averaged temperatures (that is, the mean temperature of each latitude circle) plotted as a function of latitude and altitude can therefore depict a more complete thermal structure of the atmosphere. From Figure, it may be noted that the tropopause level is highest in the tropics (30S, 30N) reaching up to about 16 km due to intense convective overturning in the equatorial region with coldest temperatures (200 K). However, its height is about 8 km over poles (absence of convective overturning) with temperatures 10 - 20 K warmer than those over the equatorial region even during perpetual darkness of winters over poles. At this level, an abrupt change in temperature (or lapse rate) may be noted both in Polar Regions and the Tropics; that is, temperatures decrease with altitude below the tropopause and rise above this level. However, such an abrupt change is not seen in the mid-latitudes, which is evident from breaks in the tropopause both in the southern and northern hemispheres. Winds with intense shear, are strongest in these regions attaining jet speeds (50 ms-1 and above) at 200 hPa over southeastern United States, Mediterranean Sea and Japan are referred to as the jet streams.

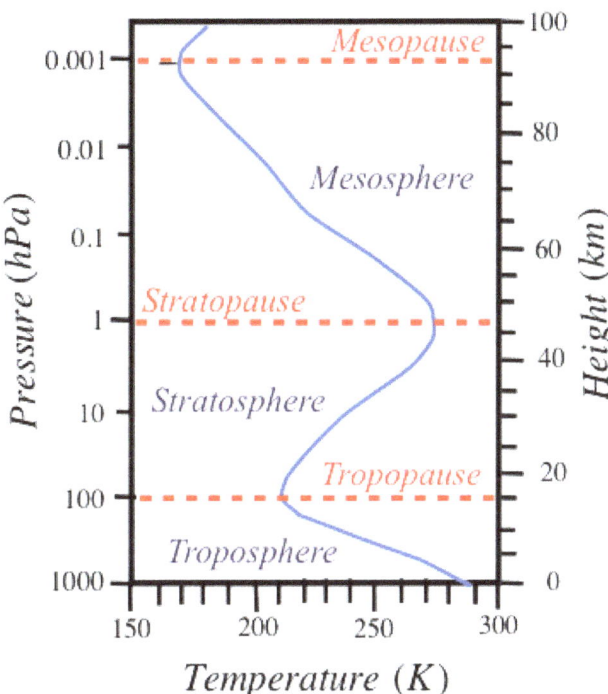

Variation of mean temperature (K) with height up to an altitude of 100 km in the atmosphere
(Fleming et al. 1990).

Temperature decreases in troposphere (lapse rate = 10K/km (dry); 6.5K/km (moist)) with height, and increases in stratosphere attaining maximum at the stratopause (~ 50 km). These two layers together contain 99% of mass of the atmosphere.

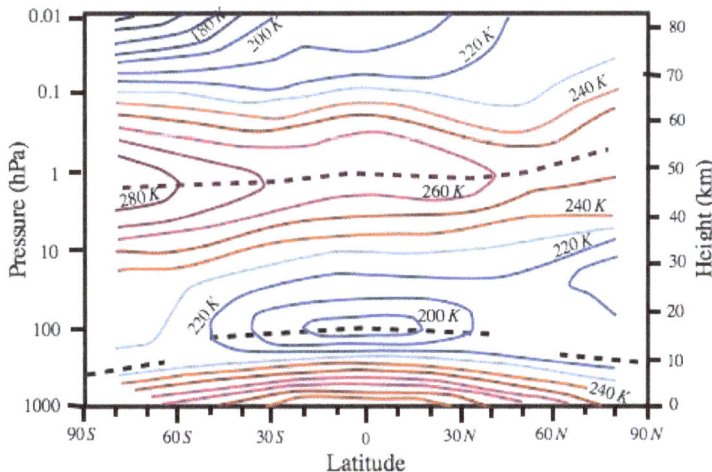

Height-latitude cross-sections of zonal-mean atmospheric temperatures (K) during northern winter. The dashed lines show tropopause (100 hPa) and stratopause (1 hPa) where lapse rate abruptly changes. Note also breaks in the tropopause at mid-latitudes. (Temperature above 80 km, refer Fleming et al. (1990)

The concentration of atmospheric constituents varies in the vertical (i.e. it is a function of altitude). For example, the distribution of zonal mean water vapour mixing ratio shows that it is very much confined to the troposphere with maximum amounts (20 g/kg) at the surface falling off rapidly with altitude in the lower troposphere

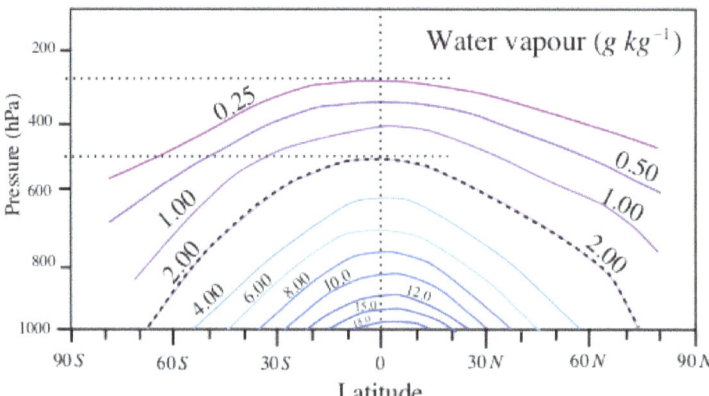

Variation of zonal mean water vapour mixing ratio with latitude and pressure. Shaded area represents 60% of the maximum value. Mixing ratios superior to 18 g/kg are at surface and 8 g/kg at about 800 hPa (~ 2 km) in tropics as found by Oort and Pexito (1983).

The water vapour mixing ratios fall off below 4 g/kg beyond 60 degree latitude poleward horizontally and above the boundary layer in the vertical apart from tropics. Exceptions are the cloudy regions where water vapour mixing ratios are high. For example, deep convective (cumulonimbus) clouds in tropics can penetrate the entire depth of the troposphere with their bases above boundary layer and cloud tops reaching up to tropopause. Thus, the largest fraction of high clouds forms in the tropics. Water vapour mostly originates over warm equatorial ocean waters due to evaporation and transported by the winds to other regions. One of the most illustrative examples of such a kind of transport phenomena is the summer monsoon, which visits the Indian subcontinent every year and remains active for about four months producing a quantum jump in its daily rainfall. Vast amounts of moisture that originate over the southern Indian Ocean (moisture source region) are transported by the monsoon current to produce copious rains over India (moisture sink region). Since water vapour is an efficient absorber of the terrestrial radiation, the patterns of outgoing longwave radiation (OLR) and water vapour match very closely. However, ozone is mainly concentrated in the stratosphere and mesosphere. The maximum ozone number density (number of molecules per cubic metre) is in the stratosphere, a stable layer often referred to as the "ozone layer" in the atmosphere.

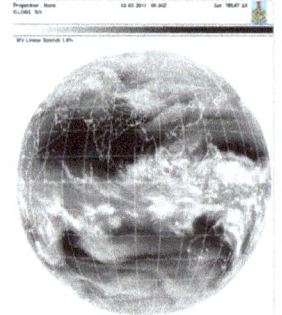

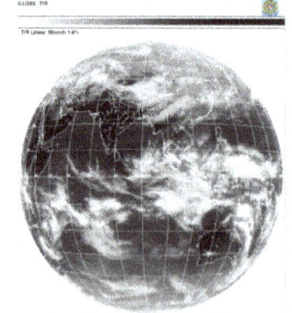

Figure 1.4a Water vapour distribution as observed from the Indian Satellite INSAT 3A on March 3, 2012 at 9.00 GMT.
(Source: India Meteorological Department)

Figure 1.4b Cloud distribution (OLR image) as observed from INSAT 3A on March 3, 2012 at 9.00 GMT.
(Source: India Meteorological Department)

Both water vapour and ozone are trace gases that are produced in some regions and destroyed elsewhere after being transported by atmospheric circulation. Ozone is constantly produced through photochemical process involving absorption of ultra-violet radiation from sun: three O_2 molecules form two O_3 molecules. Destruction of O_3 by chemicals (such aschlorofluorocarbons) creating the "ozone hole" on the Antarctica is one of the important discoveries of the last century. Higher concentrations of ozone in the polar region of the winter hemisphere suggests significant transport of O_3 due to winds, thus the equilibrium levels of O_3 concentration is the net result of its production, destruction by chemicals and transport by the winds. Ozone is also present in the mesosphere.

The ozone destroying chlorofluorocarbons (CFCs) are also significant contributors to the greenhouse effect. The anthropogenic activity releases molecules containing carbon, nitrogen and sulfur atoms in the atmosphere. These gases are released through the burning of fossil fuels and biomass, emissions from plants, and decay of plants and animals. The main scavenging mechanism for these gasses is the oxidation process where the hydroxyl radicals (HOx) play an important role. The study of chemical transformations in atmosphere has given birth to "Atmospheric Chemistry" which has now become an integral part of atmospheric and ocean sciences.

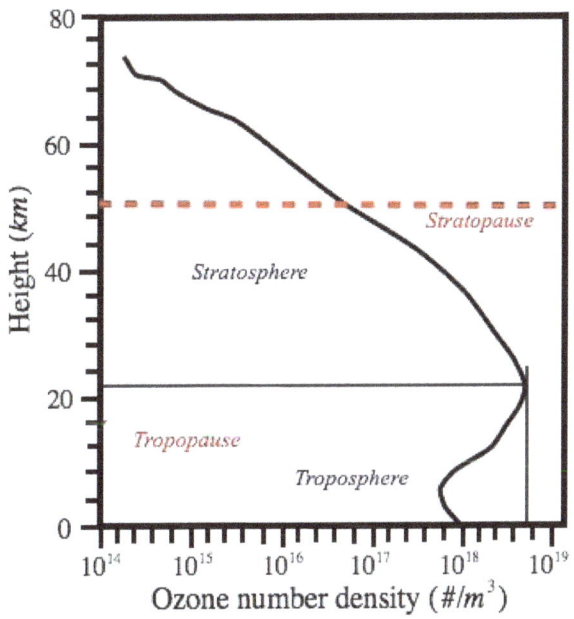

Vertical profile of ozone number density (# of molecules per cubic metre) from U.S. Standard Atmosphere (1976). Dominant levels of O3 may be noted in the stratosphere (the ozone layer) and it is present in the atmosphere up to a height of 70 km. Thus bulk of the O3 is present in the "middle atmosphere (stratosphere and mesosphere)". Ozone concentration is maximum at a height of 22 km but temperature is maximum at the stratopause level (~50 km).

Much of the information on ozone can be obtained from the observed annual cycle in the "column ozone" which gives the distribution of "total number of ozone molecules in a vertical column of atmosphere" as a function of latitude and season as shown in Figure. Column ozone is reported in Dobson Unit (DU), which is defined as the height of the column in hundredths of a millimetre if all the ozone molecules in it were compressed to a pressure of 1 atm and a temperature of 00C. Typical values of column ozone in the atmosphere are about 300 DU or 3 mm. Thus, one may note low values of column ozone (260 – 280 DU) throughout the year in tropics even though most of the ozone production is found to be in this region. Ozone amounts peak during spring at high latitudes (480 DU) in the Northern Hemisphere but it happens at middle latitudes (360 DU) in the Southern Hemisphere as seen in Figure.

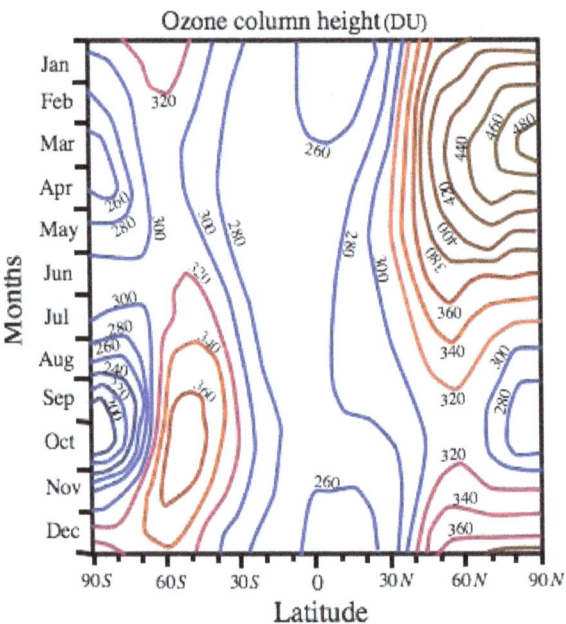

The observed annual cycle of column ozone in the 5-year climatology of Li and Shine (1995). Column ozone amounts are in Dobson Unit (DU). O3 peaks in NH high latitudes (480DU) and SH middle latitudes (360 DU) during spring. Note the "Ozone Hole" over the Antarctica with low O3 concentrations (200 DU). 1 DU = 1/100 mm of column ozone 1 DU = 2.6 ×1020 molecules O3 cm^{-2}.

Another distinctive feature of column ozone distribution could be noted during Antarctic spring (September-October) when column ozone drops to alarmingly low values over the South Pole (200 DU), a manifestation of the "Antarctic ozone hole" discovered by J.C. Farman, B.G. Gardiner and J.D. Shaklin, which was largest in September 2006. This dramatic reduction of ozone in the stratosphere over Antarctica has been found to be the result of loss of ozone, which happens due to complex chemical and physical processes, involving chlorine from chlorofluorocarbons (CFCs), on the ice particle which can form in extremely low temperature in this region. Although the ozone levels rise later in the year but the Ozone Hole over Antarctica became the subject of intense theoretical and observational studies and international efforts paved the way for phasing out the use of CFCs in accordance to the Montreal Protocol and later the Kyoto Agreement on the reduction of emission of greenhouse gases to save the planet from the potential damages of global warming.

Carbon dioxide, a major greenhouse gas of global concern, is well mixed in the atmosphere. Most of the atmospheric CO_2 (397 ppm) is used up by plant life through the photosynthesis process. Thus, oceans also absorb carbon dioxide due to the ubiquitous presence of phytoplankton in saline waters that are grazed by zooplanktons feeding fish. However, the secular increase of CO_2 in the atmosphere, evident from, would have far-reaching consequences, as mean atmospheric temperatures would also rise leading to global warming. Excessive absorption of CO_2 by oceans can, in effect, change the pH of ocean waters; this could be disastrous to ocean life because some kinds of rare fish and organisms could disappear permanently. For a systematic study of changes in the concentration of atmospheric constituents resulting from chemical transformations,it is necessary to constantly probe the chemical composition of the atmosphere.

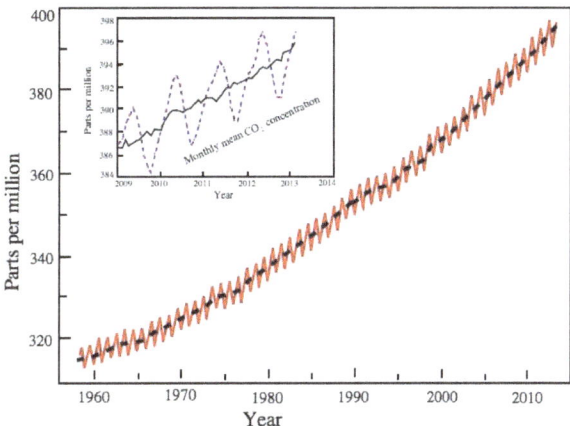

Mixing ratio of carbon dioxide over Mauna Loa, Hawaii. The mean annual
cycle of CO_2 is shown in the inset picture.
The NOAA observations show a secular rise in the levels of carbon dioxide. Note the changing slope in the rise of CO_2.
Along the x-axis the years are sown and on the ordinate CO_2 concentration in parts per million (ppm) is shown.

Composition of the Atmosphere

The composition and physical properties of the atmosphere and ocean are intimately linked to each other due to air-sea interaction at the interface. The age of the earth is estimated to be 4.6×10^9 (4.6 Ga). During the period of cooling which began some 3.8 Ga before present, oceans evolved from precipitation of transitory steam into liquid form during planetary accretion. The present day atmosphere is composed of the chemical species that are abundant in the atmosphere and though some species are present in trace amounts yet play an important role in maintaining the vertical thermal structure of the atmosphere.

The fractional concentration of the chemical constituents in the atmosphere is summarized in Table in accordance to the number of molecules or partial pressures (i.e. concentration by volume). In the present times, climate and "climate change" have become important topics of research with past and present observations including proxy data, and by using the numerical models of climate prediction. For understanding the atmosphere, we need to determine its mass and composition that is continuously changing due to the dominant anthropogenic activity. Thus, any change in the concentration of even trace gases (such as ozone) would point to the perils of industrialization and large-scale destruction of vegetation cover or warming of oceans. The state of the atmosphere in motion may however be predicted from an initial state, days and weeks ahead using the laws of nature. But, for climate and climate change, it is necessary to understand the factors that determine the average state of the atmosphere and its interaction (exchange of energy) with the other components of earth system, viz., oceans, cryosphere, biosphere and the landmass over periods of years and centuries. To address our concern about climate change, all components of the earth system have to be considered into a mathematical framework often referred to as a *climate model – a coupled land-ocean-atmosphere model*. These models are of varied complexity that even includes the changing chemistry of the air but we shall only consider the so-called simple models in these sections.

Table: Concentration by volume of atmospheric constituents; Greenhouse gases (GHG) are also indicated in the Table. Molecular weight is in units of g/mol. (Mw of air = 28.97 g mol[-1])

Chemical constituent	Molecular weight	Concentration by volume
Permanent gases:		
Nitrogen (N_2)	28.013	78%
Oxygen (O_2)	32.00	21%
Argon (Ar)	39.95	0.93%
Trace gases		
Water vapour (H_2O)	18.02	0-5% (highly variable GHG)
Carbon dioxide (CO_2)	44.01	380 ppm (GHG)
Neon (Ne)	20.18	18 ppm
Helium (He)	4.00	5 ppm
Methane (CH_4)	16.04	1750 ppm(GHG)
Krypton (Kr)	83.80	1100 ppb
Hydrogen	2.02	500 ppb
Nitrous oxide (N_2O)	56.03	300 ppb(GHG)
Ozone (O_3)	48.00	0-500 ppb(GHG)
Carbon monoxide (CO)	28.01	120 ppb
Nitrogendioxide (NO_2)	46.01	1 ppb
Sulphur dioxide (SO_2)	64.06	200 ppt

The more complete and complex models obviously consider all the components of earth system: (i) atmosphere, (ii) oceans, (iii) cryosphere, (iv) biosphere and, (v) the land mass and its orographic features. Such models that are run on ultra superfast computers housed in huge data centres, have advanced our capability not only of making extended-range weather forecasts, but also to under-standing of climate variability from the analysis of voluminous data produced in numerical simu-lations on climates of past, present and future. Oceansare constituted of large bodies of salt water covering 72% of the area of the Earth's surface, whereas the Cryosphere is comprised of water in the solid state and contributes to reflectivity of the earth surface. Antarctica and Greenland are the dominant continental ice sheets and sea ice (also called pack-ice) covers a larger area of the earth surface than the continental ice sheet. Typical sea ice thickness measures 1-3 meters and new pack ice is formed during the cold season by freezing of water.

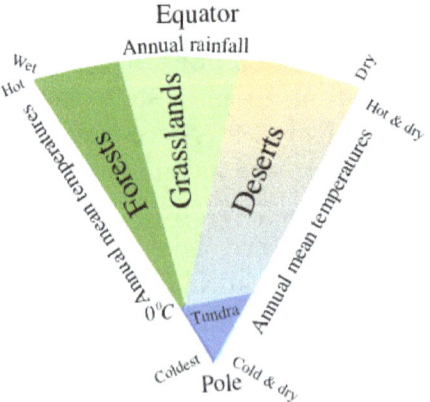

A conceptual model for displaying the land vegetation types (tundra, forest, grassland and deserts) that will grow depending on the annual rainfall and precipitation over different parts of the globe. Tundra is the treeless region with permanently frozen subsoil water (permafrost). Tundra vegetation includes small shrub like plants and lichens.

The Terrestrial Biosphere refers to the geographical distribution of forests, grasslands, tundra and deserts; their characterization mainly depends on the annual mean temperature and precipitation as shown in Figure. For example, tundra on the globe is the dominant treeless region where temperatures even in the warmest month are less than 10°C. Desert vegetation is sparsely distributed in the regions where potential evaporation (proportional to insolation i.e. solar radiation reaching the ground) exceeds precipitation. Depending upon the rainfall amounts received by a particular region, the rest of the land areas on the earth are covered by grasslands and forests. Figure clearly shows that even if water available from precipitation were same for two geographical regions but have different temperatures, then forests would cover regions where mild temperatures prevail in comparison to hot regions that could only sustain grasslands. For a thorough description of the components of the earth system, one may consult the exposition of Wallace and Hobbs (2006) on atmospheric science and other textbooks on this topic.

Vertical Structure of Oceans

Ocean is an open system like atmosphere; both ocean and atmosphere can exchange matter across their interface. But the coupled atmosphere-ocean system that they form together is a closed system separated by a common interface. Hence, thermodynamic and dynamical variations in any one component could induce changes in the climate of the whole system through exchanges at the interface. In such a coupled system, ocean is the reservoir of water in the atmosphere and maintains the hydrological cycle of the earth system. In contrast to atmosphere, ocean is opaque to all wavelengths of the solar radiation. Therefore, oceans gain heat in the equatorial latitudes by absorbing solar radiation and lose it mainly due to evaporation. Heat loss from oceans due to evaporation is however reduced in the upwelling regions (along eastern and western African coasts; western American coasts) as cold waters from the thermocline region would lower the sea surface temperatures. For example, Somali current produces strong upwelling that triggers Arabian Sea cooling during the monsoon season; but, on the contrary, if the upwelling on the Peru coast reverses, it marks the onset of a major climatic event, the El Niño, which is the ocean part of the coupled ocean-atmosphere phenomenon, El Niño Southern Oscillation (ENSO). However, cooler than normal temperatures off the coast of Peru are associated with enhanced upwelling due to stronger trades in the region; this event is referred to as La Niña and it is the opposite of El Niño in the ENSO cycle. The ocean currents also play a key role in maintaining the thermal equilibrium of the earth system by transporting heat from equatorial latitudes to Polar Regions in the upper levels and cold waters from pole to equator in deep layers. This gives rise to thermohaline circulation in the ocean on the global scale, which is often referred to as the great conveyor belt in the ocean. It takes thousands of years to complete one cycle. Is there any possibility that the conveyor belt would be switched off in future? This is an important question as it has happened during glaciation in the past. If the conveyor belt switches off, then it would lead to a mini ice age.

Another key feature of the ocean is its large thermal inertia due to seawater's large density (1027 kg m^{-3}) and greater specific heat (3986 $JK^{-1}g^{-1}$) as compared to air. Consequently, it would supposedly delay any possible CO_2 induced global warming. Besides, ocean is also a major sink of carbon dioxide. That is why the transfer of atmospheric CO_2 to deeper ocean layers by turbulent transfer makes it an important *inorganic sink* of this greenhouse gas. Moreover, marine plant life too consumes CO_2 to transform it into organic material through photosynthesis. Therefore, the abundance of phytoplankton in the ocean renders it an important *organic sink* of CO_2. Ubiquitous phytoplankton communities in the ocean are nearly responsible for 46% of the planetary photosynthesis. Much of the marine life crucially depends on phytoplankton, which receives nutrients from deeper layers of the ocean. The discussion in the preceding paragraph emphasizes the role of vertical structure of the ocean in the exchange processes and nutrient transport upwards. Like the atmosphere, the vertical structure of the ocean is also inferred from depth wise temperature distribution that is different for tropics, mid latitude and the polar regions.

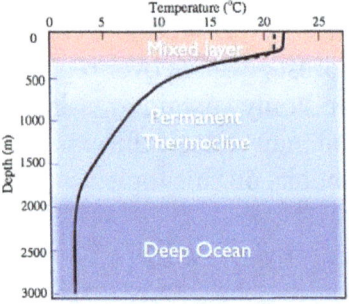

a. Temperature profiles in the tropical ocean representing its vertical structure. Thermocline is the layer of strongest vertical gradient in the temperature profile. Dashed (- - -) line is for the winter and the continuous line for the summer season.

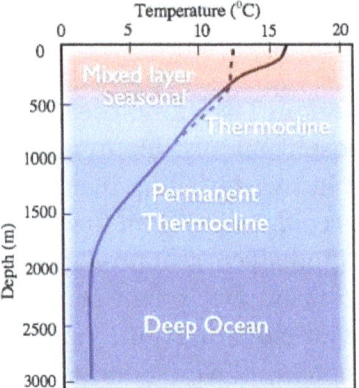

b. Temperature profiles in the mid latitudes in the ocean. Dashed (- - - -) line is for the winter and the continuous line for the summer season.

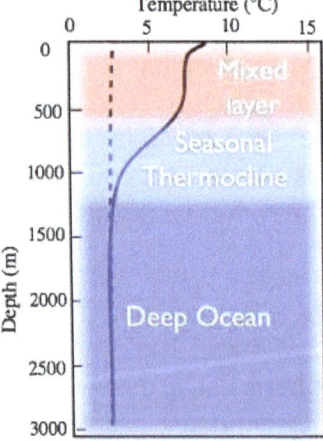

c. Ocean Tempes. in the Polar regions. Dashed (- - - -) line is for the winter and continuous line for the summer season.

From the knowledge of the temperature profiles in global ocean basins, it is convenient to divide the oceanic column into three regions (sometimes four when bottom or abyssal waters are included):

1. Surface layer (SL): The climate effects on the sea are found in the surface layer as weak gradient from equator to poles with *sigma-tee* ($\sigma_t = \rho - 1000$ kg/m^3) increasing with decreasing temperatures. Usually SL is 10-100 m but less than 200 m thick layer except in winters. It is that portion of the ocean column, which experiences seasonal changes in response to the exchange of energy with atmosphere and the absorption of solar radiation. Constant vertical overturning in the ocean mixes temperature (T) and salinity(S); therefore,both these properties are invariant with depth in the so-called "mixed layer", that is, $dq/dz=0$, $q=(T,S)$. Consequently, density which otherwise increases with decreasing temperature and increasing salinitywill have uniform vertical distribution ($\partial \rho / \partial z = 0$) in the mixed layer which is shallower in tropics than in the mid latitude ocean. Hence, it is also referred to as the isopycnal mixed layer. The depth of the 20°C isotherm in the ocean, as shown in Figure a, determines the base of the mixed layer and the corresponding depth of the 20°C isotherm as the "mixed-layer depth (MLD)". However, the 20°C isotherm *outcrops* at about 40° latitude in both the hemispheres as one may note in Figure. the other isotherms also outcrop as the surface sea temperature decrease. The outcropping of isotherms has important implication on the formation of *central waters* of the thermocline region where the role of surface wind induced mixing assumes the key role.

2. The thermocline (pycnocline) region is below the mixed layer and extends approximately up to a depth of 1000 m where temperatures are in the range of 2- 4°C. It is the region where most rapid decrease in temperature (density) occurs with depth; therefore, density increases downward and it makes the thermocline (TC) region strongly stable. The thermocline waters reach to sea surface due to wind stress forced upwelling in the ocean. The cold waters from the thermocline will reduce sea surface temperatures in the upwelling regions and of other regions where they spread by ocean currents. In this manner, the thermocline participates in the air-sea interaction in global oceans. Moreover, the upward and downward movements of thermocline (pycnocline) have important consequences for the climate especially the event like *El Niño* often referred to as the *El Niño Southern Oscillation (ENSO).*

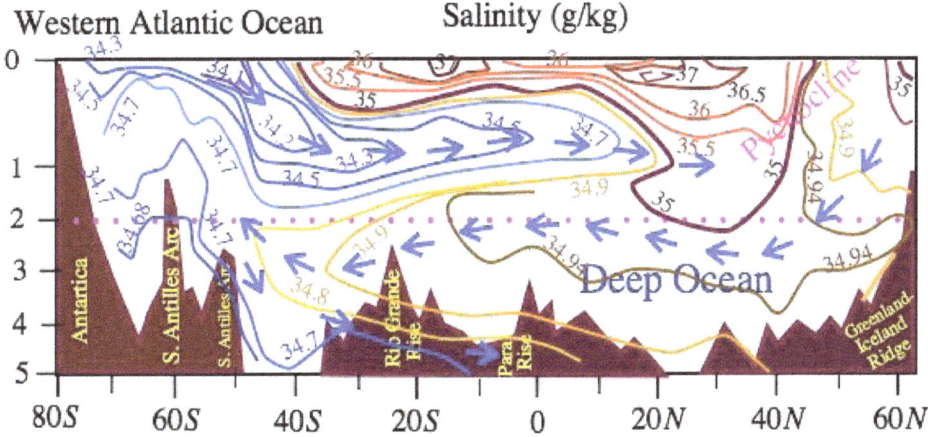

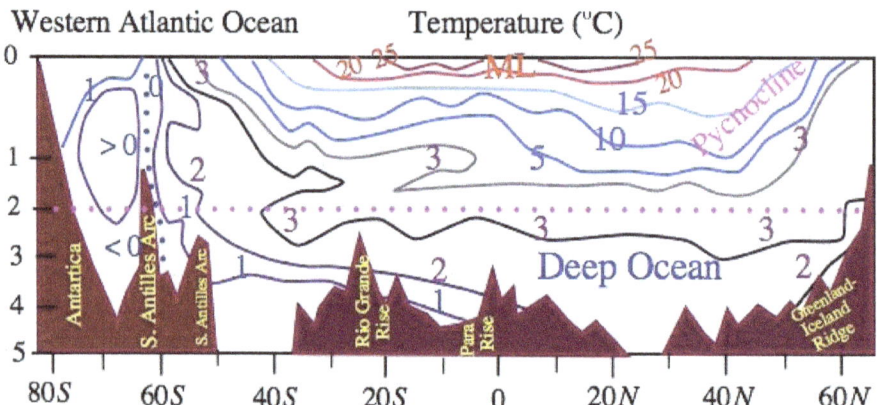

Latitude-depth temperature (°C) and salinity in the ocean representing its vertical structure. Thermocline is the layer of strongest vertical gradient in the temperature profile.

3. Deep Water (DW) layer is below the permanent thermocline and there is a gradual decrease in temperature and salinity with depth. Usually deep waters are found below a depth of 1000 meters. Deep-water temperatures are below 4°C ; and these waters being isolated from the surface by the stable thermocline, hardly interact with the atmosphere except in the regions on the surface of the ocean where deep waters are formed or destroyed. The depth of the submerged layers varies with latitude: deepest in tropics and shallowest in the high latitudes. Also, sea surface temperatures (SSTs) decrease from the equator to poles; SSTs of about 4°C and cooler would thus be found only in the high latitudes (beyond 50N or 50S). Thus, the isotherms between 0°C and 4°C characterizing deep waters, though remain submerged in tropical and midlatitude ocean depths, would outcrop at sea surface in high latitudes. Due to outcropping of isotherms, cooler surface waters would interact with deep ocean waters in high latitudes. In this manner, deep waters are formed in high latitudes or marginal seas.

Much of the information on the structure of the ocean waters is seen in the two-dimensional cross-sections of temperature and salinity as shown in Figure, which represents the structure of the Western Atlantic Ocean. The 20°C isotherm runs at a depth of about 100 m from 30S to 40N suggests that the mixed layer extends beyond tropics in the north. The expansion of mixed layer with its depth exceeding 100 m poleward beyond 30N in the North Atlantic is the result of thermal overturning due to evaporative cooling and mechanical overturning by surface winds in the storm tracks. Another interesting feature of the ocean thermal structure is seen in the thermocline region, which is below the mixed layer only in the tropics. The outcropping of isotherms of the thermocline region may be noted in the midlatitude (30°-50°). The same feature is also seen for the isotherms below 4°C , which characterize deep waters in the ocean. The latitude-depth cross-sections of salinity shows that the low salinity water from the 40S – 60S flow down below the mixed layer and a front moves towards equator at a depth between 500 m to 1 km. Another notable observation from this figure is that the thermocline is relatively shallow in the Southern Hemisphere (depth < 1 km) as compared to the Northern Hemisphere where the 4°C isotherm is running at a depth of 1.5 km.

Since the isotherms in the thermocline region south of the equator are almost straight, therefore, with reference to the equatorward moving density front below the mixed layer, low-density waters are upstream and high-density waters downstream. This sort of horizontal density variations could cause strong mixing in the region where isopycnal lines slope, which can be inferred from the analogy that isopycnal lines $d\rho/dt=0$ in the ocean are like isentropes $d\theta/dt=0$ in the atmosphere. From Figure b, one can also notice that in the north Atlantic region the heavier waters subduct to a depth of about 3 km and move southwards. The tongue of anomalously dense waters spreads beyond 40S at 3000 m depth. On traversing such huge distances at a speed of less than 1 mm/s (also difficult to measure), the dense waters from north Atlantic finally overrides much heavier waters formed in the Weddell Sea in the south. The deep waters then drift eastward to move around Antarctica and a part of which drifts northwards in the Indian Ocean and through upwelling the nutrient-rich deep cold waters find their way into shallower depths of this basin in about 2000 years.

Composition of Ocean Waters

The density of seawater depends on temperature and salinity, thus it is important to know the composition of seawater besides the temperatures. Major constituents of seawater are salts in the ionic state, which are believed to have come from the weathering of continental rocks. However, more accurate methods of analysis reveal that river waters possess potassium, magnesium, calcium and sodium ions, but they are deficient in dissolved carbon dioxide, chlorides, bromides and iodides. Hence, constituentsof seawater have different sources: potassium, magnesium, calcium and sodium ions come from continental rocks mostly transported by rivers; dissolved carbon dioxide (carbonates, bicarbonates), chlorides, bromides and iodides have originated from volcanic and hot spring discharges into the sea, and from atmospheric sources. These ions are lost constantly by sedimentation on the ocean floorand through transfer to the atmosphere and biosphere. Therefore, there is a fine geochemical balance of sources and sinks of these ions. This balance has not changed approximately for 100 Ma years from the present day composition of seawater as shown in Table and its pH must have remained constant so that the marine life could evolve in the present form. The most remarkable properties of water are: (i) maximum density of water at 4°C; (ii) expansion of water at freezing temperature; of course, (iii) salinity influences both.

The constituents of seawater allowed oceanographers to define the salt content of ocean by a single physical quantity called salinity. The chloride content of the ocean waters is easily measured, therefore, the salinity is defined in terms of chlorinity (Cl), which in turn is defined as the amount of chlorine (grams) in 1 kg of seawater with both bromine and iodine replace by chlorine. Hence salinity is defined as,

$$Salinity = 1.80655 \times [Cl]$$

The salinity is expressed in Practical Salinity Units (psu) and I psu is given by,1 psu = 1.00510 × salinity

In 2010, a new standard TEOS-10 was adopted for the calculation of the thermodynamic properties of seawater. Thus salinity is not defined in terms of psu, but defined as the true mass fraction called "Absolute Salinity (S_A)" which has units of gram per kilogram (g kg^{-1}). Thus, S_A of any seawater sample is the mass fraction of the dissolved solute in the so-called Standard Seawater (SSW) having same density as that of the given sample. Since most of the salinity data are reported and archived in Practical Salinity (S_P), the Absolute Salinity is determined from S_P by performing the following 3 steps in accordance to TEOS-10:

Step 1: Take salinity value, S_p, of the sample of the seawater;

Step 2: For the range $2 < S_p < 42$, $-2°C < T < 35°C$, estimate the Reference Salinity S_R as,

$$S_R = \frac{35.16504}{35} \times S_p \quad in\ units\ of\ g\ kg^{-1}$$

Step 3: Because S_R will not be equal to actual Absolute Salinity, they differ by a small correction factor δS_R (the salinity anomaly): $S_A = S_R + \delta S_A$

Table: Standard seawater constituents (salinity = 34.7 psu)		
Constituen tion	Mass of ion(g/kg)	Percentage of global salt
Chloride (Cl⁻)	19.215	54.96%
Sodium(Na⁺)	10.685	30.58%
Sulfate (SO$_4^-$	2.693	7.70%
Magnesium(Mg²⁺)	1.287	3.69%
Calcium (Ca²⁺)	0.410	1.17%
Potassium(K⁺)	0.396	1.13%
Bicarbonate (HCO$_3^-$)	0.142	0.41%
Bromide (Br⁻)	0.067	0.19%
Boric acid (H$_3$BO$_3$)	0.026	0.07%

There are several ways of estimating δS_A ; for example, in the absence of information one can set $\delta S_A = 0$. For an open ocean (Atlantic, Pacific, Indian, Southern Ocean, and also Baltic) there exists a global atlas of δS_A . However, one can make estimate of salinity anomaly from direct measurements of the density anomaly $\delta \rho_A$. The density anomaly is the difference between true density and that calculated using S_R as the salinity argument in the TEOS-10. Then

$$\delta S_A (g\ kg^{-1}) = \delta S_A (kg\ m^{-3}) / 0.75179$$

In many applications, it is better to use Absolute Salinity, for example, in numerical simulations.

Heating and Cooling of Ocean Waters

The ocean surface is influenced by the climate, which is apparent in the ocean by the presence of sea ice, balance of evaporation and precipitation and the distribution of salinity and temperature at the surface affecting seawater density. The gentle gradient of temperature from equator to pole suggest that there is a thermally direct convective cell driven by heat and density gradient that takes warm waters poleward at the surface and cold waters equatorward at depth. This simple conveyor belt concept has changed as a result of studies by Stommel on the mechanism of meridional exchange in the ocean that emphasizes the role of wind stress at the ocean surface in achieving planetary heat balance. The ocean climates are thus largely determined by the response of the ocean to the atmospheric forcing in which both ocean and atmosphere respond to the global solar heating as a coupled system.

The dominant processes that heat or cool the ocean are associated with radiative exchanges, evaporation and conduction from or into the ocean surface. Practically, all solar radiation reaching the sea will be absorbed in a thin layer of water, which will heat surface waters. Similarly condensation of water vapour as dew near the surface and conduction of heat from the overlaying warm atmospheric layer will heat the ocean. The ocean cooling primarily happens as a result of evaporation and conduction of heat from the ocean surface. The wind-induced evaporation is a very dominant cooling mechanism of the sea surface. For calculating the heat budget of the ocean, it is required to calculate the fluxes of sensible, latent and radiative heating to and from the ocean surface.

Formation of Different Types of Waters in Ocean Volume

Alexander von Humboldt (1814) explained that ocean waters in the tropics are cold at great depths due to sinking of cold waters in the polar latitudes. This explanation has been used to interpret the different types of waters in the ocean volume in relation to the surface effects of the marine climate in different latitudes. That is, different types of waters form at the surface continuously and sink under the climatic influences to form large masses of such water types (as distinguished from their salinity and temperature) in the volume of the ocean. As an illustration, we explain the formation of Central waters of middle latitudes in the thermocline region under the action of wind with the formation of Ekman layer. The Ekman layer dynamics will be explained later in these sections. It may be noted from Figure a that isotherms outcrop in the middle latitudes and subtropics and if under the action of wind forcing two surface water masses at different temperatures and salinity mix, then the resultant water mass will be heavier to the water masses that mixed. Such water masses will subduct also under the influence of wind along the isotherms as shown in Figure to fill the volume of the ocean. In this manner the permanent thermocline waters are formed at the surface of the ocean. Moreover, there is a constant renewal of the water masses in the interior layers with those formed at the ocean surface, which being in contact with atmosphere richin oxygen to support the marine life at depths. However, the most amazing fact about the ocean is that its *composition is constant*, yet the vertical gradient of salt concentration establishes the effectiveness of horizontal mixing processes in the ocean though vertical exchanges in the oceans are extremely weak as the lamination of water masses suggest. The *sleeping* ocean at depths, especially below the permanent thermocline (stable layer) and highly active surface ocean layers under the climatic stress of winds makes the dynamics of oceans very complicated and interesting.

The subducted waters that slide along the isotherms, are constituted of water mass with particles of same history and origin. It is also assumed that density of seawater changes happen at the surface under the combined action of radiative heating or cooling, by evaporation or precipitation, and by freezing or melting of ice. The formation of different water types in the ocean has been discussed thoroughly by W.S. von Arx (*An Introduction to Physical Oceanography, Addison-Wesley Publishing Co., 1977*).

To summarise, the surface layer water characteristics of the world ocean can be divided into three categories, viz., the Equatorial waters of mid-tropics, the Central waters of middle latitudes and the subpolar waters of the high latitudes. Figure b shows that the subpolar water tends to sink and slide equatorward as it is cooled being in contact with cold polar atmosphere. The laminates of different water masses produce a stratified structure in entire the ocean depth. The stratification is generally stable but waters in the surface layer are constantly conditioned and oxygenated being in immediate vicinity of the atmosphere in motion.

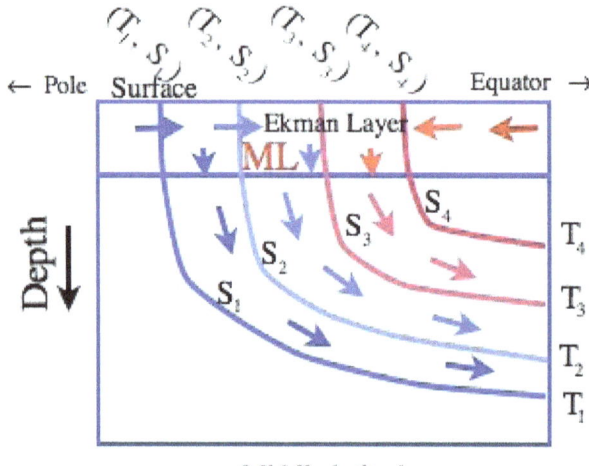

Middle latitudes

Formation of the Central waters at the surface in the Mixed Layer (ML) and their subduction in the middle latitudes. The subducted water slides along the isotherms coinciding with isopycnals to fill the volume of the ocean in the thermocline region. This renewal process is continuous and a necessary component of ocean circulation that enriches the bottom water with oxygen when the water masses come in contact with the atmosphere.

On the Evolution of Atmosphere and Ocean

The Earth and the other planets in the solar system formed approximately 4.6 billion years (4.6 Ga) ago through accretion process, which continued for about 150 million years. The oceans particularly formed at least 3.8 Ga before present due to continued precipitation as the Earth cooled. The water in the liquid form filled the ocean basins since then and it allowed life to evolve on earth throughout the geological times. The evidence of first living cell comes from the oldest known fossils of blue-green algae dated 3.5 Ga before present. The earth has undergone several glacial-interglacial cycles and the greenhouse gases in its atmospheric envelope have played a key role in the earth's climate. The first glaciation has occurred between 2.5 and 2.3 Ga, but the oceans have never frozen completely. The comparison of Earth with its neighbouring planets, Venus and Mars, is also very interesting because all the three planets have formed through the same accretion process.

References

- Jacobson, Mark Z. (June 2005). Fundamentals of Atmospheric Modeling (paperback) (2nd ed.). New York: Cambridge University Press. p. 828. ISBN 978-0-521-54865-6

- American Institute of Physics. Atmospheric General Circulation Modeling. Archived 2008-03-25 at the Wayback Machine. Retrieved on 2008-01-13

- Williamson, Fiona (2015-09-01). "Weathering the empire: meteorological research in the early British straits settlements". The British Journal for the History of Science. 48 (3): 475–492. ISSN 1474-001X. doi:10.1017/S000708741500028X

- Bluestein, H., Synoptic-Dynamic Meteorology in Midlatitudes: Principles of Kinematics and Dynamics, Vol. 1, Oxford University Press, 1992; ISBN 0-19-506267-1

- Glickman, Todd S. (June 2009). Meteorology Glossary (electronic) (2nd ed.). Cambridge, Massachusetts: American Meteorological Society. Retrieved March 10, 2014

- Edward N. Lorenz, "Deterministic non-periodic flow", Journal of the Atmospheric Sciences, vol. 20, pages 130–141 (1963)

- Rice, A. L. (1999). "The Challenger Expedition". Understanding the Oceans: Marine Science in the Wake of HMS Challenger. Routledge. pp. 27–48. ISBN 978-1-85728-705-9

- Gattuso, J.-P.; Hansson, L. (15 September 2011). Ocean Acidification. Oxford University Press. ISBN 978-0-19-959109-1. OCLC 730413873

- Office of the Federal Coordinator of Meteorology. Federal Meteorological Handbook No. 1 - Surface Weather Observations and Reports: September 2005. Retrieved on 2009-01-02

- Hamblin, Jacob Darwin (2005) Oceanographers and the Cold War: Disciples of Marine Science. University of Washington Press. ISBN 978-0-295-98482-7

- Richard J. Pasch, Mike Fiorino, and Chris Landsea. TPC/NHC'S REVIEW OF THE NCEP PRODUCTION SUITE FOR 2006. Retrieved on 2008-05-05

- Boling Guo, Daiwen Huang. Infinite-Dimensional Dynamical Systems in Atmospheric and Oceanic Science, 2014, World Scientific Publishing, ISBN 978-981-4590-37-2. Sample Chapter

- "Modern research in nuclear meteorology" (PDF). Atomic Energy. Springer New York. February 1974. doi:10.1007/BF01117823. Retrieved July 6, 2008

- Brasseur, Guy (1984). Aeronomy of the Middle Atmosphere : Chemistry and Physics of the Stratosphere and Mesosphere. Springer. pp. xi. ISBN 978-94-009-6403-7

- "Sir John Murray (1841-1914) - Founder Of Modern Oceanography". Science and Engineering at The University of Edinburgh. Retrieved 7 November 2013

- Nagy, Andrew F.; Balogh, André; Thomas E. Cravens; Mendillo, Michael; Mueller-Woodarg, Ingo (2008). Comparative Aeronomy. Springer. pp. 1–2. ISBN 978-0-387-87824-9

- "Ocean acidification". Department of Sustainability, Environment, Water, Population & Communities: Australian Antarctic Division. 28 September 2007. Retrieved 17 April 2013

- Chapman, Sydney (1960). The Thermosphere - the Earth's Outermost Atmosphere. Physics of the Upper Atmosphere. Academic Press. p. 4. ISBN 978-0-12-582050-9

Weather: An Integrated Study

Weather is the state of the atmosphere of a place. Some of the common weather phenomena are currents, wind and rain. Weather can be predicted by using computer-based models which helps in considering all the facets related to the weather. This chapter will provide an integrated understanding of weather.

Weather

Thunderstorm near Garajau, Madeira

Weather is the state of the atmosphere, to the degree that it is hot or cold, wet or dry, calm or stormy, clear or cloudy. Most weather phenomena occur in the lowest level of the atmosphere, the troposphere, just below the stratosphere. Weather refers to day-to-day temperature and precipitation activity, whereas climate is the term for the averaging of atmospheric conditions over longer periods of time. When used without qualification, "weather" is generally understood to mean the weather of Earth.

Weather is driven by air pressure, temperature and moisture differences between one place and another. These differences can occur due to the sun's angle at any particular spot, which varies with latitude. The strong temperature contrast between polar and tropical air gives rise to the largest scale atmospheric circulations: the Hadley Cell, the Ferrel Cell, the Polar Cell, and the jet stream. Weather systems in the mid-latitudes, such as extratropical cyclones, are caused by instabilities of the jet stream flow. Because the Earth's axis is tilted relative to its orbital plane, sunlight is incident at different angles at different times of the year. On Earth's surface, temperatures usually range ±40 °C (−40 °F to 100 °F) annually. Over thousands of years, changes in Earth's orbit can affect the amount and distribution of solar energy received by the Earth, thus influencing long-term climate and global climate change.

Surface temperature differences in turn cause pressure differences. Higher altitudes are cooler than lower altitudes as most atmospheric heating is due to contact with the Earth's surface while radiative losses to space are mostly constant. Weather forecasting is the application of science and technology to predict the state of the atmosphere for a future time and a given location. The Earth's weather system is a chaotic system; as a result, small changes to one part of the system can grow to have large effects on the system as a whole. Human attempts to control the weather have occurred throughout history, and there is evidence that human activities such as agriculture and industry have modified weather patterns.

Studying how the weather works on other planets has been helpful in understanding how weather works on Earth. A famous landmark in the Solar System, Jupiter's *Great Red Spot*, is an anti-cyclonic storm known to have existed for at least 300 years. However, weather is not limited to planetary bodies. A star's corona is constantly being lost to space, creating what is essentially a very thin atmosphere throughout the Solar System. The movement of mass ejected from the Sun is known as the solar wind.

Causes

Cumulus mediocris cloud surrounded by stratocumulus

On Earth, the common weather phenomena include wind, cloud, rain, snow, fog and dust storms. Less common events include natural disasters such as tornadoes, hurricanes, typhoons and ice storms. Almost all familiar weather phenomena occur in the troposphere (the lower part of the atmosphere). Weather does occur in the stratosphere and can affect weather lower down in the troposphere, but the exact mechanisms are poorly understood.

Weather occurs primarily due to air pressure, temperature and moisture differences between one place to another. These differences can occur due to the sun angle at any particular spot, which varies by latitude from the tropics. In other words, the farther from the tropics one lies, the lower the sun angle is, which causes those locations to be cooler due the spread of the sunlight over a greater surface. The strong temperature contrast between polar and tropical air gives rise to the large scale atmospheric circulation cells and the jet stream. Weather systems in the mid-latitudes, such as extratropical cyclones, are caused by instabilities of the jet stream flow. Weather systems in the tropics, such as monsoons or organized thunderstorm systems, are caused by different processes.

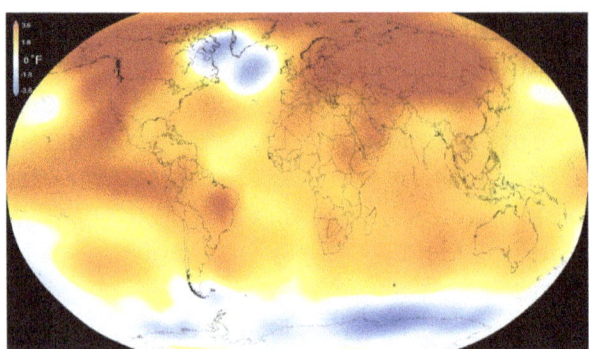

2015 – Warmest Global Year on Record (since 1880) – Colors indicate temperature anomalies
(NASA/NOAA; 20 January 2016).

Because the Earth's axis is tilted relative to its orbital plane, sunlight is incident at different angles at different times of the year. In June the Northern Hemisphere is tilted towards the sun, so at any given Northern Hemisphere latitude sunlight falls more directly on that spot than in December. This effect causes seasons. Over thousands to hundreds of thousands of years, changes in Earth's orbital parameters affect the amount and distribution of solar energy received by the Earth and influence long-term climate.

The uneven solar heating (the formation of zones of temperature and moisture gradients, or frontogenesis) can also be due to the weather itself in the form of cloudiness and precipitation. Higher altitudes are typically cooler than lower altitudes, which the result of higher surface temperature and radiational heating, which produces the adiabatic lapse rate. In some situations, the temperature actually increases with height. This phenomenon is known as an inversion and can cause mountaintops to be warmer than the valleys below. Inversions can lead to the formation of fog and often act as a cap that suppresses thunderstorm development. On local scales, temperature differences can occur because different surfaces (such as oceans, forests, ice sheets, or man-made objects) have differing physical characteristics such as reflectivity, roughness, or moisture content.

Surface temperature differences in turn cause pressure differences. A hot surface warms the air above it causing it to expand and lower the density and the resulting surface air pressure. The resulting horizontal pressure gradient moves the air from higher to lower pressure regions, creating a wind, and the Earth's rotation then causes deflection of this air flow due to the Coriolis effect. The simple systems thus formed can then display emergent behaviour to produce more complex systems and thus other weather phenomena. Large scale examples include the Hadley cell while a smaller scale example would be coastal breezes.

The atmosphere is a chaotic system, as a result, small changes to one part of the system can grow to have large effects on the system as a whole. This makes it difficult to accurately predict weather more than a few days in advance, though weather forecasters are continually working to extend this limit through the scientific study of weather, meteorology. It is theoretically impossible to make useful day-to-day predictions more than about two weeks ahead, imposing an upper limit to potential for improved prediction skill.

Shaping the Planet Earth

Weather is one of the fundamental processes that shape the Earth. The process of weathering breaks down the rocks and soils into smaller fragments and then into their constituent substances. During rains precipitation, the water droplets absorb and dissolve carbon dioxide from the surrounding air. This causes the rainwater to be slightly acidic, which aids the erosive properties of water. The released sediment and chemicals are then free to take part in chemical reactions that can affect the surface further (such as acid rain), and sodium and chloride ions (salt) deposited in the seas/oceans. The sediment may reform in time and by geological forces into other rocks and soils. In this way, weather plays a major role in erosion of the surface.

Global Weather Video for Year 2015

EUMETSAT created "A Year in Weather 2015" a narrated video of the earth's weather photographed from weather satellites for the entire year 2015. Geostationary satellite photographs from EUMETSAT, the Japan Meteorological Agency and the National Oceanic and Atmospheric Administration were assembled to show weather changing on earth for 365 days in a time lapse video.

Major wind and pressure systems and Related Weather

Region	Name	Pressure	Surface Winds	Weather
Equator (0°)	Doldrums (ITCZ) (equatorial low)	Low	Light, variable winds	Cloudiness, abundant precipitation in all seasons; breeding ground for hurricanes. Relatively low sea-surface salinity because of high rainfall relative to evaporation
0°–30°N and S	Trade winds (easterlies)	-	Northeast in Northern Hemisphere; Southeast in Southern Hemisphere	Summer wet, winter dry; pathway for tropical disturbances
30°N and S	Horse latitudes	High	Light, variable winds	Little cloudiness; dry in all seasons. Relatively high sea-surface salinity because of high evaporation relative to precipitation
30°–60°N and S	Prevailing Westerlies	-	Southwest in Northern Hemisphere; Northwest in Southern Hemisphere	Winter wet, summer dry; pathway for subtropical high and low pressure
60°N and S	Polar front	Low	Variable	Stormy, cloudy weather zone; ample precipitation in all seasons
60°–90°N and S	Polar easterlies	-	Northeast in Northern Hemisphere; Southeast in Southern Hemisphere	Cold polar air with very low temperatures
90°N and S	Poles	High	Southerly in Northern Hemisphere; Northerly in Southern Hemisphere	Cold, dry air; sparse precipitation in all seasons

Effect on Humans

Weather, seen from an anthropological perspective, is something all humans in the world constantly experience through their senses, at least while being outside. There are socially and scientifically constructed understandings of what weather is, what makes it change, the effect it has on humans in different situations, etc. Therefore, weather is something people often communicate about.

Effects on Populations

New Orleans, Louisiana, after being struck by Hurricane Katrina. Katrina was a Category 3 hurricane
when it struck although it had been a category 5 hurricane in the Gulf of Mexico.

Weather has played a large and sometimes direct part in human history. Aside from climatic
changes that have caused the gradual drift of populations (for example the desertification of the
Middle East, and the formation of land bridges during glacial periods), extreme weather events
have caused smaller scale population movements and intruded directly in historical events. One
such event is the saving of Japan from invasion by the Mongol fleet of Kublai Khan by the Kami-
kaze winds in 1281. French claims to Florida came to an end in 1565 when a hurricane destroyed
the French fleet, allowing Spain to conquer Fort Caroline. More recently, Hurricane Katrina redis-
tributed over one million people from the central Gulf coast elsewhere across the United States,
becoming the largest diaspora in the history of the United States.

The Little Ice Age caused crop failures and famines in Europe. The 1690s saw the worst famine in
France since the Middle Ages. Finland suffered a severe famine in 1696–1697, during which about
one-third of the Finnish population died.

Forecasting

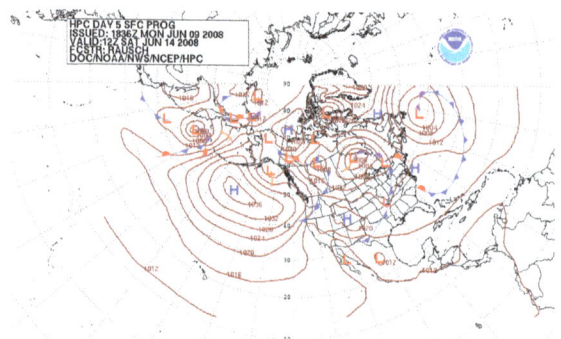

Forecast of surface pressures five days into the future for the north Pacific, North America,
and north Atlantic Ocean as on 9 June 2008

Weather forecasting is the application of science and technology to predict the state of the atmosphere for a future time and a given location. Human beings have attempted to predict the weather
informally for millennia, and formally since at least the nineteenth century. Weather forecasts are
made by collecting quantitative data about the current state of the atmosphere and using scientific
understanding of atmospheric processes to project how the atmosphere will evolve.

Once an all-human endeavor based mainly upon changes in barometric pressure, current weather conditions, and sky condition, forecast models are now used to determine future conditions. Human input is still required to pick the best possible forecast model to base the forecast upon, which involves pattern recognition skills, teleconnections, knowledge of model performance, and knowledge of model biases. The chaotic nature of the atmosphere, the massive computational power required to solve the equations that describe the atmosphere, error involved in measuring the initial conditions, and an incomplete understanding of atmospheric processes mean that forecasts become less accurate as the difference in current time and the time for which the forecast is being made (the *range* of the forecast) increases. The use of ensembles and model consensus helps to narrow the error and pick the most likely outcome.

There are a variety of end users to weather forecasts. Weather warnings are important forecasts because they are used to protect life and property. Forecasts based on temperature and precipitation are important to agriculture, and therefore to commodity traders within stock markets. Temperature forecasts are used by utility companies to estimate demand over coming days. On an everyday basis, people use weather forecasts to determine what to wear on a given day. Since outdoor activities are severely curtailed by heavy rain, snow and the wind chill, forecasts can be used to plan activities around these events, and to plan ahead and survive them.

Modification

The aspiration to control the weather is evident throughout human history: from ancient rituals intended to bring rain for crops to the U.S. Military Operation Popeye, an attempt to disrupt supply lines by lengthening the North Vietnamese monsoon. The most successful attempts at influencing weather involve cloud seeding; they include the fog- and low stratus dispersion techniques employed by major airports, techniques used to increase winter precipitation over mountains, and techniques to suppress hail. A recent example of weather control was China's preparation for the 2008 Summer Olympic Games. China shot 1,104 rain dispersal rockets from 21 sites in the city of Beijing in an effort to keep rain away from the opening ceremony of the games on 8 August 2008. Guo Hu, head of the Beijing Municipal Meteorological Bureau (BMB), confirmed the success of the operation with 100 millimeters falling in Baoding City of Hebei Province, to the southwest and Beijing's Fangshan District recording a rainfall of 25 millimeters.

Whereas there is inconclusive evidence for these techniques' efficacy, there is extensive evidence that human activity such as agriculture and industry results in inadvertent weather modification:

- Acid rain, caused by industrial emission of sulfur dioxide and nitrogen oxides into the atmosphere, adversely affects freshwater lakes, vegetation, and structures.

- Anthropogenic pollutants reduce air quality and visibility.

- Climate change caused by human activities that emit greenhouse gases into the air is expected to affect the frequency of extreme weather events such as drought, extreme temperatures, flooding, high winds, and severe storms.

- Heat, generated by large metropolitan areas have been shown to minutely affect nearby weather, even at distances as far as 1,600 kilometres (990 mi).

The effects of inadvertent weather modification may pose serious threats to many aspects of civilization, including ecosystems, natural resources, food and fiber production, economic development, and human health.

Microscale Meteorology

Microscale meteorology is the study of short-lived atmospheric phenomena smaller than mesoscale, about 1 km or less. These two branches of meteorology are sometimes grouped together as "mesoscale and microscale meteorology" (MMM) and together study all phenomena smaller than synoptic scale; that is they study features generally too small to be depicted on a weather map. These include small and generally fleeting cloud "puffs" and other small cloud features.

Extremes on Earth

Early morning sunshine over Bratislava, Slovakia

The same area, just three hours later, after light snowfall

On Earth, temperatures usually range ±40 °C (100 °F to −40 °F) annually. The range of climates and latitudes across the planet can offer extremes of temperature outside this range. The coldest air temperature ever recorded on Earth is −89.2 °C (−128.6 °F), at Vostok Station, Antarctica on 21 July 1983. The hottest air temperature ever recorded was 57.7 °C (135.9 °F) at 'Aziziya, Libya, on 13 September 1922, but that reading is queried. The highest recorded average annual temperature was 34.4 °C (93.9 °F) at Dallol, Ethiopia. The coldest recorded average annual temperature was −55.1 °C (−67.2 °F) at Vostok Station, Antarctica.

The coldest average annual temperature in a permanently inhabited location is at Eureka, Nunavut, in Canada, where the annual average temperature is −19.7 °C (−3.5 °F).

Extraterrestrial within the Solar System

Studying how the weather works on other planets has been seen as helpful in understanding how it works on Earth. Weather on other planets follows many of the same physical principles as weather on Earth, but occurs on different scales and in atmospheres having different chemical composition. The Cassini–Huygens mission to Titan discovered clouds formed from methane or ethane which deposit rain composed of liquid methane and other organic compounds. Earth's atmosphere includes six latitudinal circulation zones, three in each hemisphere. In contrast, Jupiter's banded appearance shows many such zones, Titan has a single jet stream near the 50th parallel north latitude, and Venus has a single jet near the equator.

Jupiter's Great Red Spot in 1979

One of the most famous landmarks in the Solar System, Jupiter's *Great Red Spot*, is an anticyclonic storm known to have existed for at least 300 years. On other gas giants, the lack of a surface allows the wind to reach enormous speeds: gusts of up to 600 metres per second (about 2,100 km/h or 1,300 mph) have been measured on the planet Neptune. This has created a puzzle for planetary scientists. The weather is ultimately created by solar energy and the amount of energy received by Neptune is only about $\frac{1}{900}$ of that received by Earth, yet the intensity of weather phenomena on Neptune is far greater than on Earth. The strongest planetary winds discovered so far are on the extrasolar planet HD 189733 b, which is thought to have easterly winds moving at more than 9,600 kilometres per hour (6,000 mph).

Space Weather

Aurora Borealis

Weather is not limited to planetary bodies. Like all stars, the sun's corona is constantly being lost to space, creating what is essentially a very thin atmosphere throughout the Solar System. The movement of mass ejected from the Sun is known as the solar wind. Inconsistencies in this wind and larger events on the surface of the star, such as coronal mass ejections, form a system that has features analogous to conventional weather systems (such as pressure and wind) and is generally known as space weather. Coronal mass ejections have been tracked as far out in the solar system as Saturn. The activity of this system can affect planetary atmospheres and occasionally surfaces. The interaction of the solar wind with the terrestrial atmosphere can produce spectacular aurorae, and can play havoc with electrically sensitive systems such as electricity grids and radio signals.

Space Weather

Aurora australis observed by *Discovery*, May 1991

Space weather is a branch of space physics and aeronomy concerned with the time varying conditions within the Solar System, including the solar wind, emphasizing the space surrounding the Earth, including conditions in the magnetosphere, ionosphere and thermosphere. Space weather is distinct from the terrestrial weather of the Earth's atmosphere (troposphere and stratosphere). The science of space weather is focused on fundamental research and practical applications. The term *space weather* was first used in the 1950s and came into common usage in the 1990s.

History

For many centuries, the effects of space weather were noticed but not understood. Displays of auroral light have long been observed at high latitudes.

Genesis

In 1724, George Graham reported that the needle of a magnetic compass was regularly deflected from magnetic north over the course of each day. This effect was eventually attributed to overhead electric currents flowing in the ionosphere and magnetosphere by Balfour Stewart in 1882, and confirmed by Arthur Schuster in 1889 from analysis of magnetic observatory data.

In 1852, astronomer and British major general Edward Sabine showed that the probability of the occurrence of magnetic storms on Earth was correlated with the number of sunspots, thus demonstrating a novel solar-terrestrial interaction. In 1859, a great magnetic storm caused brilliant auroral displays and disrupted global telegraph operations. Richard Carrington correctly connected the storm with a solar flare that he had observed the day before in the vicinity of a large sunspot group—thus demonstrating that specific solar events could affect the Earth.

Kristian Birkeland explained the physics of aurora by creating artificial aurora in his laboratory and predicted the solar wind. The introduction of radio revealed that periods of extreme static or noise occurred. Severe radar jamming during a large solar event in 1942 led to the discovery of solar radio bursts (radio waves which cover a broad frequency range created by a solar flare), another aspect of space weather.

Twentieth Century

In the 20th century the interest in space weather expanded as military and commercial systems came to depend on systems affected by space weather. Communications satellites are a vital part of global commerce. Weather satellite systems provide information about terrestrial weather. The signals from satellites of the Global Positioning System (GPS) are used in a wide variety of applications. Space weather phenomena can interfere with or damage these satellites or interfere with the radio signals with which they operate. Space weather phenomena can cause damaging surges in long distance transmission lines and expose passengers and crew of aircraft travel to radiation, especially on polar routes.

The International Geophysical Year (IGY), increased research into space weather. Ground-based data obtained during IGY demonstrated that the aurora occurred in an *auroral oval*, a permanent region of luminescence 15 to 25 degrees in latitude from the magnetic poles and 5 to 20 degrees wide. In 1958, the Explorer I satellite discovered the Van Allen belts, regions of radiation particles trapped by the Earth's magnetic field. In January 1959, the Soviet satellite Luna 1 first directly observed the solar wind and measured its strength.

In 1969, INJUN-5 (a.k.a. Explorer 40) made the first direct observation of the electric field impressed on the Earth's high latitude ionosphere by the solar wind. In the early 1970s, Triad data demonstrated that permanent electric currents flowed between the auroral oval and the magnetosphere.

The term space weather came into usage in the 1990s along with the belief that space's impact on human systems demanded a more coordinated research and application framework.

US National Space Weather Program

The purpose of the US National Space Weather Program is to focus research on the needs of the affected commercial and military communities, to connect the research and user communities, to create coordination between operational data centers and to better define user community needs.

The concept was turned into an action plan in 2000, an implementation plan in 2002, an assessment in 2006 and a revised strategic plan in 2010. A revised action plan was scheduled to be released in 2011 followed by a revised implementation plan in 2012.

One part of the National Space Weather Program is to show users that space weather affects their business. Private companies now acknowledge space weather "is a real risk for today's businesses".

Phenomena

Within the solar system, space weather is influenced by the solar wind and the interplanetary magnetic field (IMF) carried by the solar wind plasma. A variety of physical phenomena are associated with space weather, including geomagnetic storms and substorms, energization of the Van Allen radiation belts, ionospheric disturbances and scintillation of satellite-to-ground radio signals and long-range radar signals, aurora and geomagnetically induced currents at Earth's surface. Coronal mass ejections and their associated shock waves are also important drivers of space weather as they can compress the magnetosphere and trigger geomagnetic storms. Solar energetic particles (SEP), accelerated by coronal mass ejections or solar flares, are also an important driver of space weather as they can damage electronics onboard spacecraft (e.g. Galaxy 15 failure), and threaten the lives of astronauts.

Effects

Spacecraft Electronics

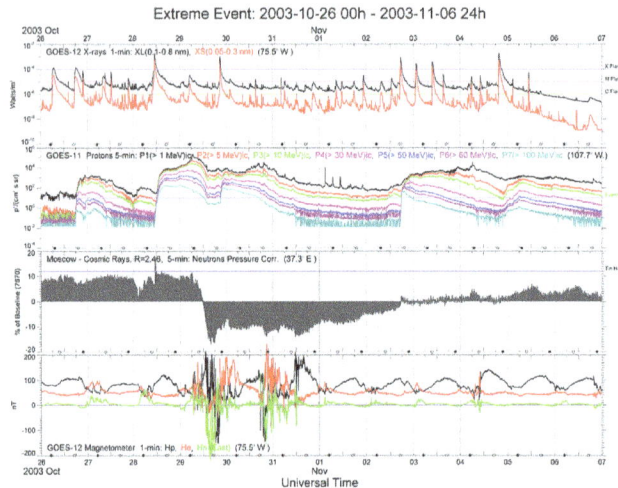

GOES-11 and GOES-12 monitored space weather conditions during the October 2003 solar activity.

Some spacecraft failures can be directly attributed to space weather; many more are thought to have a space weather component. For example, 46 of the 70 failures reported in 2003 occurred during the October 2003 geomagnetic storm. The two most common adverse space weather effects on spacecraft are radiation damage and spacecraft charging.

Radiation (high energy particles) passes through the skin of the spacecraft and into the electronic components. In most cases the radiation causes an erroneous signal or changes one bit in memory of a spacecraft's electronics (single event upsets). In a few cases, the radiation destroys a section of the electronics (single-event latchup).

Spacecraft charging is the accumulation of an electrostatic charge on a non-conducting material on the spacecraft's surface by low energy particles. If enough charge is built-up, a discharge (spark) occurs. This can cause an erroneous signal to be detected and acted on by the spacecraft computer. A recent study indicates that spacecraft charging is the predominant space weather effect on spacecraft in geosynchronous orbit.

Spacecraft Orbit Changes

The orbits of spacecraft in low Earth orbit (LEO) decay to lower and lower altitudes due to the resistance from the friction between the spacecraft's surface (*i.e.*, drag) and the outer layer of the Earth's atmosphere (a.k.a. the thermosphere and exosphere). Eventually, a LEO spacecraft falls out of orbit and towards the Earth's surface. Many spacecraft launched in the past couple of decades have the ability to fire a small rocket to manage their orbits. The rocket can increase altitude to extend lifetime, to direct the reentry towards a particular (marine) site, or route the satellite to avoid collision with other spacecraft. Such maneuvers require precise information about the orbit. A geomagnetic storm can cause an orbit change over a couple of days that otherwise would occur over a year or more. The geomagnetic storm adds heat to the thermosphere, causing the thermosphere to expand and rise, increasing the drag on spacecraft. The 2009 satellite collision between the Iridium 33 and Cosmos 2251 demonstrated the importance of having precise knowledge of all objects in orbit. Iridium 33 had the capability to maneuver out of the path of Cosmos 2251 and could have evaded the crash, if a credible collision prediction had been available.

Humans in Space

The exposure of a human body to ionizing radiation has the same harmful effects whether the source of the radiation is a medical X-ray machine, a nuclear power plant or radiation in space. The degree of the harmful effect depends on the length of exposure and the radiation's energy density. The ever-present radiation belts extend down to the altitude of manned spacecraft such as the International Space Station (ISS) and the Space Shuttle, but the amount of exposure is within the acceptable lifetime exposure limit under normal conditions. During a major space weather event that includes an SEP burst, the flux can increase by orders of magnitude. Areas within ISS provide shielding that can keep the total dose within safe limits. For the Space Shuttle, such an event would have required immediate mission termination.

Ground Systems

Art inspired from the concept of space weather

...als

...waves in the same manner that water in a swimming pool bends visi-
...gh which such waves travel is disturbed, the light image or radio

information is distorted and can become unrecognizable. The degree of distortion (scintillation) of a radio wave by the ionosphere depends on the signal frequency. Radio signals in the VHF band (30 to 300 MHz) can be distorted beyond recognition by a disturbed ionosphere. Radio signals in the UHF band (300 MHz to 3 GHz) transit a disturbed ionosphere, but a receiver may not be able to keep locked to the carrier frequency. GPS uses signals at 1575.42 MHz (L1) and 1227.6 MHz (L2) that can be distorted by a disturbed ionosphere. Space weather events that corrupt GPS signals can significantly impact society. For example, the Wide Area Augmentation System (WAAS) operated by the US Federal Aviation Administration is used as a navigation tool for North American commercial aviation. It is disabled by every major space weather event. Outages can range from minutes to days. Major space weather events can push the disturbed polar ionosphere 10° to 30° of latitude toward the equator and can cause large ionospheric gradients (changes in density over distance of hundreds of km) at mid and low latitude. Both of these factors can distort GPS signals.

Long-distance Radio Signals

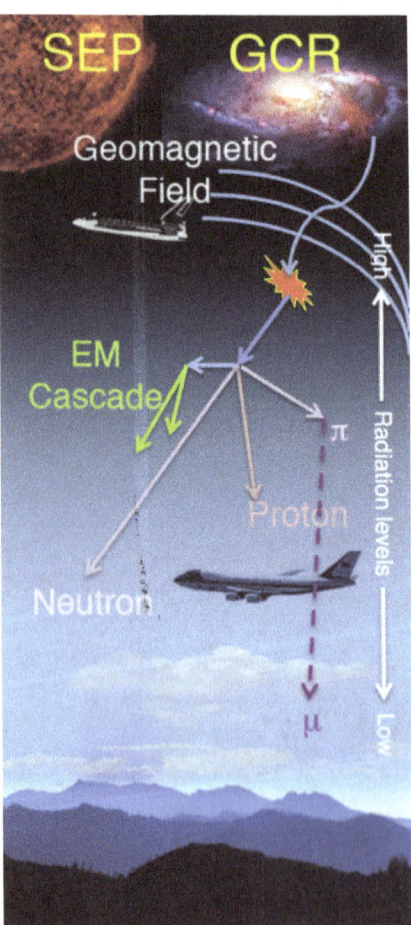

All passengers in commercial aircraft flying above 26,000 feet will typically experience some
exposure in this aviation radiation environment.

Radio wave in the HF band (3 to 30 MHz) (also known as the shortwave band) are reflected by the ionosphere. Since the ground also reflects HF waves, a signal can be transmitted around the curvature of the Earth beyond the line of sight. During the 20th century, HF communications was the only method for a ship or aircraft far from land or a base station to communicate. The s of systems such as Iridium brought other methods of communications, but HF remains critical for vessels that do not carry the newer equipment and as a critical backup system for others. Space weather events can create irregularities in the ionosphere that scatter HF signals instead of reflecting them, preventing HF communications. At auroral and polar latitudes, small space weather events that occur frequently disrupt HF communications. At mid-latitudes, HF communications are disrupted by solar radio bursts, by X-rays from solar flares (which enhance and disturb the ionospheric D-layer) and by TEC enhancements and irregularities during major geomagnetic storms.

Transpolar airline routes are particularly sensitive to space weather, in part because Federal Aviation Regulations require reliable communication over the entire flight. Diverting such a flight is estimated to cost about $100,000.

Humans in Commercial Aviation

The magnetosphere guides cosmic ray and solar energetic particles to polar latitudes, while high energy charged particles enter the mesosphere, stratosphere and troposphere. These energetic particles at the top of the atmosphere shatter atmospheric atoms and molecules, creating harmful lower energy particles that penetrate deep into the atmosphere and create measurable radiation. All aircraft flying above 8 km (26,200 feet) altitude are exposed to these particles. The dose exposure is greater in polar regions than at mid-latitude and equatorial regions. Many commercial aircraft fly over the polar region. When a space weather event causes radiation exposure to exceed the safe level set by aviation authorities, the aircraft's flight path is diverted.

While the most significant, but highly unlikely, health consequences to atmospheric radiation exposure include death from cancer due to long-term exposure, many lifestyle-degrading and career-impacting cancer forms can also occur. A cancer diagnosis can have significant career impact for a commercial pilot. A cancer diagnosis can ground a pilot temporarily or permanently. International guidelines from the International Commission on Radiological Protection (ICRP) have been developed to mitigate this statistical risk. The ICRP recommends effective dose limits of a 5-year average of 20 mSv per year with no more than 50 mSv in a single year for non-pregnant, occupationally exposed persons, and 1 mSv per year for the general public. Radiation dose limits are not engineering limits. In the U.S., they are treated as an upper limit of acceptability and not a regulatory limit.

Measurements of the radiation environment at commercial aircraft altitudes above 8 km (26,000 ft) have historically been done by instruments that record the data on board where the data are then processed later on the ground. However, a system of real-time radiation measurements onboard aircraft has been developed through the NASA Automated Radiation Measurements for Aerospace Safety (ARMAS) program. ARMAS has flown hundreds of flights since 2013, mostly on research aircraft, and sent the data to the ground through Iridium satellite links. The eventual goal of these types of measurements is to data assimilate them into physics-based global radiation models, e.g., NASA's Nowcast of Atmospheric Ionizing Radiation System (NAIRAS), so as to provide the weather of the radiation environment rather than the climatology.

Ground-induced Electric Fields

Magnetic storm activity can induce geoelectric fields in the Earth's conducting lithosphere. Corresponding voltage differentials can find their way into electric power grids through ground connections, driving uncontrolled electric currents that interfere with grid operation, damage transformers, trip protective relays and sometimes cause blackouts. This complicated chain of causes and effects was demonstrated during the magnetic storm of March 1989, which caused the complete collapse of the Hydro-Québec electric-power grid in Canada, temporarily leaving nine million people without electricity. The possible occurrence of an even more intense storm led to operational standards intended to mitigate induction-hazard risks, while reinsurance companies commissioned revised risk assessments.

Geophysical Exploration

Air- and ship-borne magnetic surveys can be affected by rapid magnetic field variations during geomagnetic storms. Such storms cause data interpretation problems because the space-weather-related magnetic field changes are similar in magnitude to those of the sub-surface crustal magnetic field in the survey area. Accurate geomagnetic storm warnings, including an assessment of storm magnitude and duration allows for an economic use of survey equipment.

Geophysics and Hydrocarbon Production

For economic and other reasons, oil and gas production often involves horizontal drilling of well paths many kilometers from a single wellhead. Accuracy requirements are strict, due to target size – reservoirs may only be a few tens to hundreds of meters across – and safety, because of the proximity of other boreholes. The most accurate gyroscopic method is expensive, since it can stop drilling for hours. An alternative is to use a magnetic survey, which enables measurement while drilling (MWD). Near real-time magnetic data can be used to correct drilling direction. Magnetic data and space weather forecasts can help to clarify unknown sources of drilling error.

Terrestrial Weather

The amount of energy entering the troposphere and stratosphere from space weather phenomena is trivial compared to the solar insolation in the visible and infra-red portions of the solar electromagnetic spectrum. However, some linkage between the 11-year sunspot cycle and the Earth's climate has been claimed. For example, the Maunder minimum, a 70-year period almost devoid of sunspots, allegedly correlated to a cooler climate. One suggestion for the linkage between space and terrestrial weather is that changes in cosmic ray flux cause changes in the amount of cloud formation. Another suggestion is that variations in the EUV flux subtly influence existing drivers of the climate and tip the balance between El Niño/La Niña events. However, a linkage between space weather and the climate has not been demonstrated conclusively.

Observation

Observation of space weather is done both for scientific research and for applications. Scientific observation has evolved with the state of knowledge, while application-related observation expanded with the ability to exploit such data.

Ground-based

Space weather is monitored at ground level by observing changes in the Earth's magnetic field over periods of seconds to days, by observing the surface of the Sun and by observing radio noise created in the Sun's atmosphere.

The Sunspot Number (SSN) is the number of sunspots on the Sun's photosphere in visible light on the side of the Sun visible to an Earth observer. The number and total area of sunspots are related to the brightness of the Sun in the extreme ultraviolet (EUV) and X-ray portions of the solar spectrum and to solar activity such as solar flares and coronal mass ejections (CMEs).

10.7 cm radio flux (F10.7) is a measurement of RF emissions from the Sun and is approximately correlated with the solar EUV flux. Since this RF emission is easily obtained from the ground and EUV flux is not, this value has been measured and disseminated continuously since 1947. The world standard measurements are made by the Dominion Radio Astrophysical Observatory at Penticton, B.C., Canada and reported once a day at local noon in solar flux units ($10^{-22}W\cdot m^{-2}\cdot Hz^{-1}$). F10.7 is archived by the National Geophysical Data Center.

Fundamental space weather monitoring data are provided by ground-based magnetometers and magnetic observatories. Magnetic storms were first discovered by ground-based measurement of occasional magnetic disturbance. Ground magnetometer data provide real-time situational awareness for post-event analysis. Magnetic observatories have been in continuous operations for decades to centuries, providing data to inform studies of long-term changes in space climatology.

Dst index is an estimate of the magnetic field change at the Earth's magnetic equator due to a ring of electric current at and just earthward of the geosynchronous orbit. The index is based on data from four ground-based magnetic observatories between 21° and 33° magnetic latitude during a one-hour period. Stations closer to the magnetic equator are not used due to ionospheric effects. The Dst index is compiled and archived by the World Data Center for Geomagnetism, Kyoto.

Kp/ap Index: 'a' is an index created from the geomagnetic disturbance at one mid-latitude (40° to 50° latitude) geomagnetic observatory during a 3-hour period. 'K' is the quasi-logarithmic counterpart of the 'a' index. Kp and ap are the average of K and an over 13 geomagnetic observatories to represent planetary-wide geomagnetic disturbances. The Kp/ap index indicates both geomagnetic storms and substorms (auroral disturbance). Kp/ap is available from 1932 onward.

AE index is compiled from geomagnetic disturbances at 12 geomagnetic observatories in and near the auroral zones and is recorded at 1-minute intervals. The public AE index is available with a lag of two to three days that limits its utility for space weather applications. The AE index indicates the intensity of geomagnetic substorms except during a major geomagnetic storm when the auroral zones expand equatorward from the observatories.

Radio noise bursts are reported by the Radio Solar Telescope Network to the U.S. Air Force and to NOAA. The radio bursts are associated with solar flare plasma that interacts with the ambient solar atmosphere.

The Sun's photosphere is observed continuously for activity that can be the precursors to solar flares and CMEs. The Global Oscillation Network Group (GONG) project monitors both the surface and the interior of the Sun by using helioseismology, the study of sound waves propagating through the Sun and observed as ripples on the solar surface. GONG can detect sunspot groups on the far side of the Sun. This ability has recently been verified by visual observations from the

STEREO spacecraft.

Neutron monitors on the ground indirectly monitor cosmic rays from the Sun and galactic sources. When cosmic rays interact with the atmosphere, atomic interactions occur that cause a shower of lower energy particles to descend into the atmosphere and to ground level. The presence of cosmic rays in the near-Earth space environment can be detected by monitoring high energy neutrons at ground level. Small fluxes of cosmic rays are present continuously. Large fluxes are produced by the Sun during events related to energetic solar flares.

Total Electron Content (TEC) is a measure of the ionosphere over a given location. TEC is the number of electrons in a column one meter square from the base of the ionosphere (approximately 90 km altitude) to the top of the ionosphere (approximately 1000 km altitude). Many TEC measurements are made by monitoring the two frequencies transmitted by GPS spacecraft. Presently GPS TEC is monitored and distributed in real time from more than 360 stations maintained by agencies in many countries.

Geoeffectiveness is a measure of how strongly space weather magnetic fields, such as coronal mass ejections, couple with the Earth's magnetic field. This is determined by the direction of the magnetic field held within the plasma that originates from the Sun. New techniques measuring Faraday Rotation in radio waves are in development to measure field direction.

Satellite-based

A host of research spacecraft have explored space weather. The Orbiting Geophysical Observatory series were among the first spacecraft with the mission of analyzing the space environment. Recent spacecraft include the NASA-ESA Solar-Terrestrial Relations Observatory (STEREO) pair of spacecraft launched in 2006 into solar orbit and the Van Allen Probes, launched in 2012 into a highly elliptical Earth-orbit. The two STEREO spacecraft drift away from the Earth by about 22° per year, one leading and the other trailing the Earth in its orbit. Together they compile information about the solar surface and atmosphere in three dimensions. The Van Allen probes record detailed information about the radiation belts, geomagnetic storms and the relationship between the two.

Some spacecraft with other primary missions have carried auxiliary instruments for solar observation. Among the earliest such spacecraft were the Applications Technology Satellite (ATS) series at GEO that were precursors to the modern Geostationary Operational Environmental Satellite (GOES) weather satellite and many communication satellites. The ATS spacecraft carried environmental particle sensors as auxiliary payloads and had their navigational magnetic field sensor used for sensing the environment.

Many of the early instruments were research spacecraft that were re-purposed for space weather applications. One of the first of these was the IMP-8 (Interplanetary Monitoring Platform) It orbited the Earth at 35 Earth Radii and observed the solar wind for two-thirds of its 12-day orbits from 1973 to 2006. Since the solar wind carries disturbances that affect the magnetosphere and ionosphere, IMP-8 demonstrated the utility of continuous solar wind monitoring. IMP-8 was followed by ISEE-3, which was placed near the L_1 Sun-Earth Lagrangian point, 235 Earth radii above the surface (about 1.5 million km, or 924,000 miles) and continuously monitored the solar wind from 1978 to 1982. The next spacecraft to monitor the solar wind at the L_1 point was WIND from 1994 to

1998. After April 1998, the WIND spacecraft orbit was changed to circle the Earth and occasionally pass the L_1 point. The NASA Advanced Composition Explorer (ACE) has monitored the solar wind at the L_1 point from 1997 to present.

In addition to monitoring the solar wind, monitoring the Sun is important to space weather. Because the solar EUV cannot be monitored from the ground, the joint NASA-ESA Solar and Heliospheric Observatory (SOHO) spacecraft was launched and has provided solar EUV images beginning in 1995. SOHO is a main source of near-real time solar data for both research and space weather prediction and inspired the STEREO mission. The Yohkoh spacecraft at LEO observed the Sun from 1991 to 2001 in the X-ray portion of the solar spectrum and was useful for both research and space weather prediction. Data from Yohkoh inspired the Solar X-ray Imager on GOES.

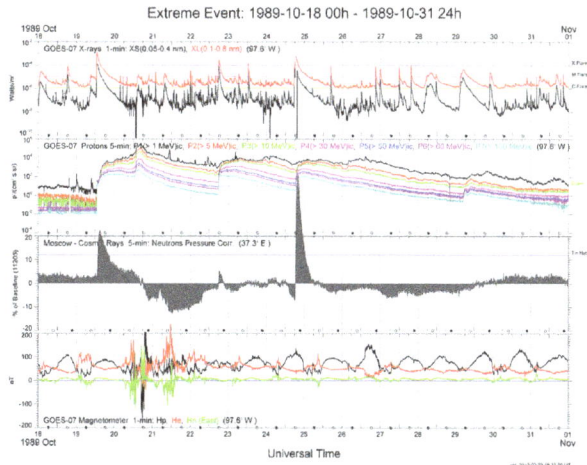

GOES-7 monitors space weather conditions during the October 1989 solar activity resulted in a Forbush Decrease, Ground Level Enhancements, and many satellite anomalies.

Spacecraft with instruments whose primary purpose is to provide data for space weather predictions and applications include the Geostationary Operational Environmental Satellite (GOES) series of spacecraft, the POES series, the DMSP series, and the Meteosat series. The GOES spacecraft have carried an X-ray sensor (XRS) which measures the flux from the whole solar disk in two bands – 0.05 to 0.4 nm and 0.1 to 0.8 nm – since 1974, an X-ray imager (SXI) since 2004, a magnetometer which measures the distortions of the Earth's magnetic field due to space weather, a whole disk EUV sensor since 2004, and particle sensors (EPS/HEPAD) which measure ions and electrons in the energy range of 50 keV to 500 MeV. Starting sometime after 2015, the GOES-R generation of GOES spacecraft will replace the SXI with a solar EUV image (SUVI) similar to the one on SOHO and STEREO and the particle sensor will be augmented with a component to extend the energy range down to 30 eV.

The Deep Space Climate Observatory (DSCOVR) satellite is a NOAA Earth observation and space weather satellite that launched in February 2015. Among its features is advance warning of coronal mass ejections.

Models

Space weather models are simulations of the space weather environment. Models use sets of mathematical equations to describe physical processes.

These models take a limited data set and attempt to describe all or part of the space weather environment in or to predict how weather evolves over time. Early models were heuristic; i.e., they did not directly employ physics. These models take less resources than their more sophisticated descendants.

Later models use physics to account for as many phenomena as possible. No model can yet reliably predict the environment from the surface of the Sun to the bottom of the Earth's ionosphere. Space weather models differ from meteorological models in that the amount of input is vastly smaller.

A significant portion of space weather model research and development in the past two decades has been done as part of the Geospace Environmental Model (GEM) program of the National Science Foundation. The two major modeling centers are the Center for Space Environment Modeling (CSEM) and the Center for Integrated Space weather Modeling (CISM). The Community Coordinated Modeling Center (CCMC) at the NASA Goddard Space Flight Center is a facility for coordinating the development and testing of research models, for improving and preparing models for use in space weather prediction and application.

Modeling techniques include (a) magnetohydrodynamics, in which the environment is treated as a fluid, (b) particle in cell, in which non-fluid interactions are handled within a cell and then cells are connected to describe the environment, (c) first principles, in which physical processes are in balance (or equilibrium) with one another, (d) semi-static modeling, in which a statistical or empirical relationship is described, or a combination of multiple methods.

Commercial Space Weather Development

During the first decade of the 21st Century, a commercial sector emerged that engaged in space weather, serving agencies, commercial and consumers. Space weather providers are mostly small companies that provide space weather data, models, derivative products and service distribution.

The commercial sector includes scientific and engineering researchers as well as users. Activities are primarily directed toward the impacts of space weather upon technology. These include, for example:

- Atmospheric drag on LEO satellites caused by energy inputs into the thermosphere from solar UV, FUV, Lyman-alpha, EUV, XUV and X-ray photons as well as by charged particle precipitation and Joule heating at high latitudes;

- Surface and internal charging from increased energetic particle fluxes, leading to effects such as discharges, single event upsets and latch-up, on LEO to GEO satellites;

- Disrupted GPS signals caused by ionospheric scintillation leading to increased uncertainty in navigation systems such as aviation's Wide Area Augmentation System (WAAS);

- Lost HF, UHF and L-band radio communications due to ionosphere scintillation, solar flares and geomagnetic storms;

- Increased radiation to human tissue and avionics from galactic cosmic rays and SEP, especially during large solar flares, at altitudes above 8 km;

- Increased inaccuracy in surveying and oil/gas exploration that uses the Earth's main magnetic field when it is disturbed by geomagnetic storms;

- Loss of power transmission from GIC surges in the electrical power grid and transformer shutdowns during large geomagnetic storms.

Many of these disturbances result in societal impacts that account for a significant part of the national GDP.

The concept of a Space Weather Economic Innovation Zone stems from the work of the American Commercial Space Weather Association (ACSWA) in 2015. The establishment of this economic innovation zone would encourage expanded economic activity developing applications to manage the risks space weather and would encourage broader research activities related to space weather by universities. The formation of a Space Weather Economic Innovation Zone would: encourage U.S. business investment in space weather services and products; support U.S. business innovation in space weather services and products; require U.S. government purchases of U.S. built commercial hardware, software, and associated products and services where no suitable government capability pre-exists; promote U.S. built commercial hardware, software, and associated products and services to international partners; designate U.S. built commercial hardware, services, and products as "Space Weather Economic Innovation Zone" activities; and track U.S. built commercial hardware, services, and products as Space Weather Economic Innovation Zone contributions within agency reports. In 2015 HR1561 provided groundwork where social and environmental impacts from a Space Weather Economic Innovation Zone could be far-reaching. U.S. built commercial hardware, services, and products can prevent economic disruptions and loss-of-life from societal risks resulting from space weather and an innovation zone would encourage the research and investment needed for this success. In 2016, the Space Weather Research and Forecasting Act (S. 2817) would potentially build on that legacy.

American Commercial Space Weather Association

On April 29, 2010, the commercial space weather community created the American Commercial Space Weather Association (ACSWA) an industry association. ACSWA promotes space weather risk mitigation for national infrastructure, economic strength and national security. It seeks to:

- provide quality space weather data and services to help mitigate risks to technology;

- provide advisory services to government agencies;

- provide guidance on the best task division between commercial providers and government agencies;

- represent the interests of commercial providers;

- represent commercial capabilities in the national and international arena;

- develop best-practices.

Understanding Weather

The atmospheric general circulation and its variance – produced by the embedded movements of moisture laden air masses, constant solar heating, in homogeneity in the earth's surface characteristics due to oceans, land, solid water mass distribution (snow and icepack), forest and deserts – have been thoroughly studied by the scientists and the savants of different epochs who have advanced our knowledge of weather, climate and their changes at various time scales. Clouds, rain and thunderstorms, clear skies, low pressure centres, cyclones in the oceans produce weather conditions that impact our daily life, agriculture, power and industry. In simple terms, weather is a manifestation of changes in air motions and distribution of rains forming from the interaction of water vapour with solar radiation. Rightly, the logo of the *India Meteorological Department (IMD)* is emblazoned with the *Sanskrit Sloka* saying *"Ādityāya jāyatā vrishtē (Rains emerge from The Sun)"*. The humankind has thus always craved for (and successfully accomplished) a deep understanding of the exchange processes (or of the physical laws) that are responsible for such weather events mainly driven by solar heating and dynamical exchanges, which are sometimes devastating to life and property. The planet earth receives energy from the Sun, and the partition of this energy happens in such a manner that there is a complete balance between the energy received and the energy used up by various terrestrial and atmospheric processes. Moreover, in the budget calculation it is necessary to consider the spherical shape of the planet earth.

It is now a common knowledge that accurate weather can be predicted using the physical laws of nature. The understanding of weather thus relates to predicting three-dimensional spatial and temporal changes that result from heating of the atmosphere, moisture distribution and air motions on the global and local scales. Continuous observations from land and space-borne platforms on weather, first by national services; and subsequently, dissemination of such data on the global network, viz., the World Weather Watch (WWW) of the World Meteorological Organization (WMO) have enabled the major forecast centres to produce regularly short-, medium- and long-range reliable weather forecasts from numerical models which use vast computer resources. The *"art of weather forecasting"* has received appreciation widely in recent years mainly because satellites put in space for constant weather watch produce enormous volumes of data of uniform quality that all go into models to accomplish successful numerical weather predictions. From a statistics of different platform data used in the forecast system of European Centre for Medium-range Weather Forecasts (ECMWF), it is estimated that more than 500,000 daily data points are assimilated in the analysis, which is regularly used in producing an operational 10-day (medium-range) high accuracy useful forecast. The majority of the data used in this prediction are from remote-sensing platforms. Further, the data usage at other major prediction centres is of the same order.

We may recall that the term *weather* refers to a *synoptic state* of the atmosphere that is experienced by a location on the earth and it is uniquely described (or determined) by temperature, humidity, rainfall; wind speed and cloud cover which continuously evolve from an interplay of meteorological processes. Whereas *climate* refers to an *average state* of the atmosphere over longer time periods and it was practically referred to as an entity independent of time. The perception has changed since and the 30-year average fields of meteorological variables are now used to represent long-term conditions and delineation of variations in the climatic averages that are necessary for *climate change* studies. Shortest time scales are observed in atmospheric variables, and ocean variables show variations on longer time scales ranging from months to thousands of years. However, the events like cyclones in the ocean induce strong air-sea interaction, which results in strong ocean surface cooling arising due to strong cooling and the lifting of the thermocline locally.

We shall now describe some of the laws, which led to the understanding of meteorological and oceanographic phenomena. One such law in meteorology was established in 1857, the year of revolt by The Great Indian Sepoy Mangal Pandey. The Dutch meteorologist C.H.D. Buys-Ballot formulated this meteorological law empirically, and it is known after his name. The Buys-Ballot's law is extensively used in the analysis of weather charts and synoptic forecasting. However, William Ferrel (1854) already derived theoretically the relationship between wind direction and the pressure distribution. Ferrel's theoretical relationship impeccably distinguishes the laboratory flows from those dominated by earth's rotation. Consider the case of a fluid flow through a conduit as an example of the laboratory flow. The movement of fluid in the conduit is down the pressure gradient; that is, the directions of velocity and of the *negative* pressure gradient coincide because the fluid flows in the tube from high pressure to low pressure. But in the geophysical fluid flows dominated by rotation of the earth, the direction of flow is perpendicular to the direction of the pressure gradient.

Buys-Ballot's Law

It describes the relationship between the wind direction and the pressure distribution. It states that *if you stand in the direction of wind blowing from your back in the Northern Hemisphere, then the low pressure is located on the left and the high pressure on the right.* In the Southern Hemisphere, *the low pressure is located on the right and the high pressure on the left while looking downstream.* This illustrates that the direction of the wind is perpendicular to pressure gradient force, which in turn implies that *wind blows parallel to the isobars.* However, in the equatorial region the law is not applicable due to weak Coriolis effect. In effect, it is true for the free atmosphere (above 1.5 km or 850 hPa), but near to the surface the wind turn towards low pressure due to frictional effects of the ground. Thus the law modifies near the surface in the following manner.

If you stand in Northern Hemisphere with your face looking downstream (i.e. wind is blowing from your back), *and turn clockwise by an angle of about 30°,* the lower pressure will be to your left and the high pressure on your right; in Southern Hemisphere, if you stand with your back to the wind (i.e. wind is blowing from your back), and *turn counter clockwise by about 30°,* the lower pressure will be to your right and the high pressure on your left.

Since meteorology deals with atmospheric motions, the wind systems over any part of globe could be interpreted in the light of the Buys-Ballot's Law. This law thus helps in understanding the weather systems and their movement from meteorological (synoptic) charts of different pressure levels. The *synoptic charts* depict the state of the atmosphere at *synoptic hours* which are produced daily by weather services of all countries for 00 GMT, 06 GMT, 12 GMT and 18 GMT (the synoptic hours).

Geostrophic Flow

In a circular pipe the flow is down the pressure gradient, i.e. fluid flows from high pressure to low pressure. *On a rotating planet the Coriolis force (CF) balances the pressure gradient force (PGF).* The Coriolis force (apparent force arises due to earth's rotation) is always directed to the right relative to the direction of the geostrophic flow in the northern hemisphere, and to the left relative to the direction of flow in the southern hemisphere. That is, Coriolis force is always perpendicular to the direction of the geostrophic flow and wind blows parallel to isobar, which is also the essence of the Buys-Ballot Law. Coriolis force balancing the pressure gradient force, one may observe in the free atmosphere above 850 hPa (1500 m) and in the interior of the ocean below the Ekman layer of depth 30 m. Indeed there is an Ekman layer at the bottom of the ocean also, which is about 35 m thick. So, between these two Ekman layers in the ocean, currents are geostrophic. Therefore, geostrophic currents are frictionless. In oceanography, currents are always expressed with reference to the oceanic floor (rotating with the earth) where current velocity is zero and effect of rotation is accounted for in the balances of forces.

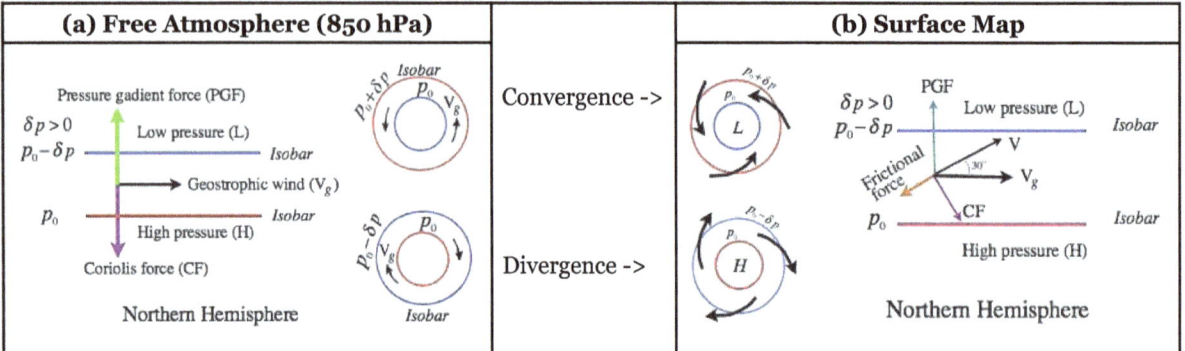

(a) Wind direction at 850 hPa under geostrophic balance: wind blows parallel to isobars; (b) Turning of the wind at surface under the action of frictional force. Note that balance of frictional force, pressure gradient force (PGF) and Coriolis force (CF) happens in such a manner that the Coriolis force is always perpendicular to the direction of flow and always acts on the right to the flow direction in the northern hemisphere. In Northern Hemisphere: Surface map shows winds spiraling clockwise out from a high pressure region (divergence); spiraling counter clockwise into low pressure (convergence). In Southern Hemisphere: the sense of wind is opposite; that is, clockwise rotation of wind in low pressure and counter clockwise rotation of wind in high pressure.

However, when friction of the earth surface comes into play, force balance happens in such a manner that Coriolis force is always perpendicular to the deflected wind blowing at an angle of 30° relative to the lines of isobars. Thus, surface friction will engender cross isobaric flows. Moreover in these circumstances, weather would evolve due to frictional convergence, which may lead to development of weather in the low pressure areas. In case if the surface pressure continues to fall with a steep rate due to this forced convergence then one may expect thunder storm activity in the region of low pressure. In a *low* at surface, convergence of wind at the surface moves air parcels upward (shown by the bold arrow in Figure a and finally divergence would take place in the upper atmosphere. On the other hand, sinking air motion occurs in high pressure regions due to convergence of mass in the upper atmosphere which finally forces a divergent flow at the surface. Thus, on a weather chart, a couplet of a low and a high could easily be noted which facilitates weather prognosis. From the configurations given in Figure, it should not be difficult for the student to identify the *centre of action:* one dominated by the weather associated with cloud and rain, and the other that will produce fair weather with good sunshine. It may also be remarked that Fridtjof Nansen (1898) first worked out the balance of wind stress, Coriolis force and frictional drag to explain the movement of iceberg at an angle 20°-40° to the direction of wind which he noticed during his historic dash in the year 1895-96 to North Pole in his ship named *Fram*.

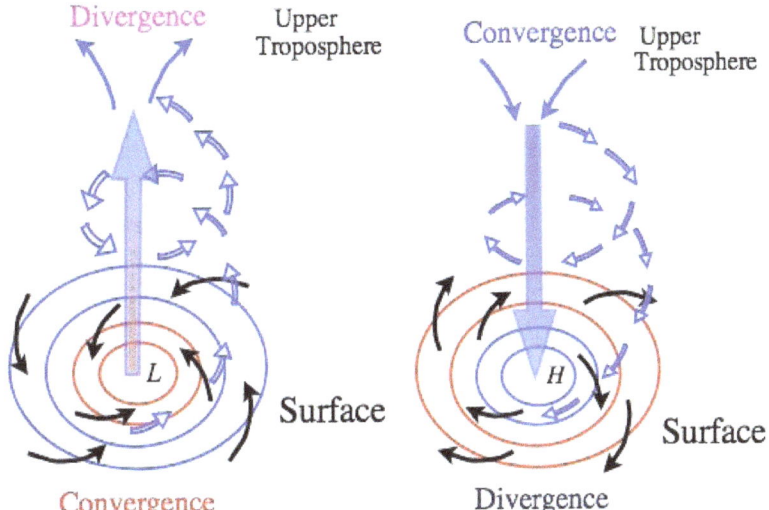

(a). Low level convergence in a low pressure region (L) at the surface; (b) convergence aloft over a surface high (H). The upper level divergence over the low (L) produces cloudy conditions and upper level convergence over the surface high (H) will give rise to clear weather.

Forced by winds, strong western boundary currents are the important components of the coastal circulation, which also give rise to upwelling in those regions. Prominent among them are the *Gulf Stream* in the North Atlantic, summertime *Somali Current* in the Indian Ocean, *Kuroshio Current* off the coast of Japan, *Brazil Current* in the Southern Atlantic, *Agulhas Current* in the South Indian Ocean, and the *East Australian Current* in the Pacific. These currents are *not geostrophic* currents but they are forced by the winds such that the costal boundaries are on west to the direction of current, which is to the *left* with respect to the direction of current in the northern hemisphere, but to the *right* in the southern hemisphere.

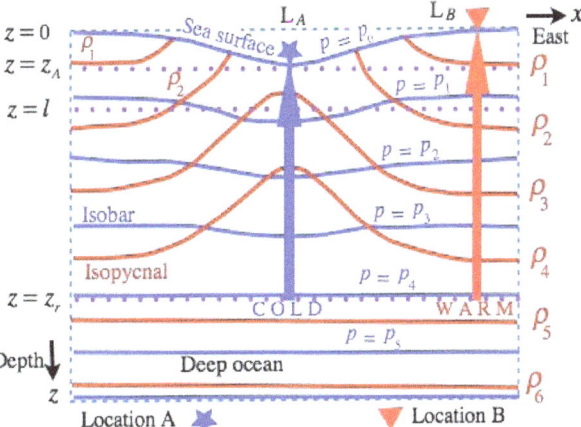

Distribution of isobars and isopycnals at any depth above z=z_r level. The water at L_A is denser than that at L_B. The steric effect, arising due to contraction or expansion of water columns, is manifested in sea surface height variations.

The disposition of isobars and isopycnals in the ocean in the northern hemisphere where geostrophic flow prevails is shown in Figure with two reference points viz., Location A (LA) and Location B (LB) in the ocean. Moreover, the height of warm (cold) water column is longer (shorter) due to expansion (contraction) relative to that of the surrounding cold (warm) water. Therefore, the water at *LA* in Figure is denser being cold than the water at LB that is warm. This is evident in the figure as the isobars are inclined upward toward LB with reference to the *LA* location; but isopycnals are inclined downward towards LB. The density changes in the seawater give rise to the *steric effect* that is manifested by variations of free surface levels as seen in Figure. Thus, the steric effect is the contraction and expansion of the water columns in the oceans, which arise from the anomalies of temperature and salinity. At reference depth $z=z_r$, pressures at locations *LA* and *LB* are same. Hence from the reference level $z=z_r$, the column height of LB has to be longer in comparison to the column height of water at LA location when measured from the reference level $z=z_r$. Note, $p_o < p_1 < p_2$ and $P_o < P_1 < P_{2...}$ though at deeper depths the density lines coincide with pressure lines (condition for barotropic structure). Further, the turbulent eddies in wind at the interface would produce eddies on the sea surface, which would, in turn, give rise to steric effects locally. As a consequence, surface water will be locally set in motion due to momentum transfer and mixing of the momentum in a vertical layer will subsequently happen. This sort of *pumping* of momentum due to the action of wind at the air-sea interface gives rise to a distinct flow in top layer known as the *Ekman layer flow* which has been discussed in what follows.

Ekman Layer Flow

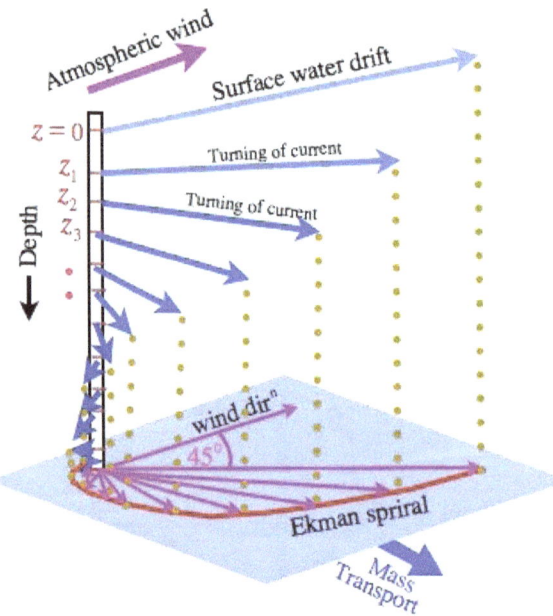

The Ekman Spiral: the direction of the surface current is at an angle of 45° to the direction of wind. The currents in the layer rotate clockwise with depth and reversal of currents may be observed below a depth that can be derived mathematically.

In the Ekman layer, friction and the Coriolis force balance and the resulting flow has spiraling velocity pattern that decays with depth. The depth of the Ekman layer varies in the range 45 - 300 m in the ocean but observed depth is much lower than the theoretical value. There is an Ekman layer at the bottom also. And between these two layers, the currents are geostrophic. The famous oceanographer V.W. Ekman developed the theory of the laminar Ekman layer in 1902 to find the surface drift current forced by the winds. The mathematical details of the solution are not given here but the graphical description given in Figure shows that currents constantly change direction. The average direction of the mass transport is 90° to the direction of the wind on the right. *That is, mass transport is perpendicular to the direction of wind and on the right relative to the direction of wind in the northern hemisphere*. In the southern hemisphere, the sense of the mass transport will be opposite. The ideal Ekman spiral is seldom observed in the atmosphere or in the ocean. However, the dynamics of the Ekman layer (i.e. the velocity pattern in the vertical) depends largely on the vertical profile of turbulent eddy viscosity.

Thermal Wind

William Ferrel found that winds are strongest when isobars are closest together and the atmosphere is in hydrostatic and geostrophic equilibrium. This fact can be exploited to calculate the strength of upper air winds using the surface pressure and temperature measurements. Geostrophic wind must have vertical shear in presence of horizontal temperature gradient; *shear in the geostrophic wind is referred to as the thermal wind in meteorology*. The thermal wind relationship is often used to check the consistency of the observations; thereby it is also serving as a diagnostic tool for observed wind and temperature fields. *Thermal wind blows parallel to isotherms (lines of constant thickness) with warm air to the right facing downstream in the Northern Hemisphere*. The *thermal wind relation* can be derived from equations that describe the geostrophic motion (characterized by small *Rossby number*) and the hydrostatic balance. For geostrophic flows both in ocean and atmosphere, inertia force is smaller than the Coriolis force; so the *Rossby number (Ro)* is small. *Ro* is defined with characteristic velocity (U) and length scale (L) as

$$R_0 = \frac{Inertia\,force}{Coriolis\,force}; \;or\; R_0 = \frac{(pu\partial u / \partial x)}{\rho\,fu} = \frac{U^2 / L}{fU} = \frac{U}{fL} \dots\dots\dots(2)$$

For geostrophic flows, Ro ~ 0.1, hence the time dependent terms in the equations of motion could be neglected; and the *geostrophic balance* is defined by eqs. (2) as

$$-fv_g = -\frac{1}{\rho}\frac{\partial p}{\partial x}\dots\dots\dots\dots(2a)$$

$$fu_g = -\frac{1}{\rho}\frac{\partial p}{\partial y}\dots\dots\dots\dots(2b)$$

In eqs. (2), u_g and v_g are the components of *geostrophic velocity* (v_g) in the *eastward (x)* and *northward (y)* directions, f=2Ω sinφ is the Coriolis parameter with Ω as the rotation frequency of earth and φ is the latitude; ρ is the *density* of the fluid. Also note that *Coriolis parameter* varies with latitude. The *hydrostatic balance* in the vertical (z) direction for both atmosphere and ocean is given by (3a), but particularly for atmosphere in the form (3b). Thus, we have

$$\frac{\partial p}{\partial z} = -gp\dots\dots\dots(2.3a)$$

$$\frac{\partial \Phi}{\partial \rho} = -\frac{1}{\rho} = -\alpha = -\frac{RT}{P}\;(using\;p = \rho RT\;for\;atmosphere)\dots\dots\dots(3b)$$

In (3b) $\Phi = gz$ is the *geopotential*. On differentiating both sides in eqs. (2) with respect to z and replacing $\partial p/\partial z$ using (3a), one obtains following equations

$$\frac{\partial v_g}{\partial z} = -\frac{g}{f\rho_0}\frac{\partial \rho}{\partial x} \quad \text{................(4a)}$$

$$\frac{\partial u_g}{\partial z} = \frac{g}{f\rho_0}\frac{\partial \rho}{\partial y} \quad \text{......................(4b)}$$

In eqs. (4a), ρ_o is the reference density, which simplifies the terms on the right. The *thermal wind relation* for the ocean is defined by the eqs. (4b) with density ρ. But in atmosphere temperature T has more relevance (than density ρ) in visualizing the consequence of the thermal wind relation. Using simple calculus, one can obtain the thermal wind equation (in vector form) for the geostrophic flow in the atmosphere as

$$\frac{\partial V_g}{\partial (\ln p)} = -\frac{R}{f}k\times\nabla_p T \quad (R \text{ is the gas cons}\tan t)$$

$$\text{.....................................(5)}$$

The eqn. (5) gives indeed the *shear* in V_g. In shear flows velocity increases or decreases with height. Integrating (5) with respect to ln p, we get the following relation for the thermal wind V_T in the atmosphere

$$V_T = V_g(p_2) - V_g(p_1) = -\frac{R}{f}\int_{p_1}^{p_2} k\times\nabla_p T d(\ln p)$$

$$\text{.....................................(6)}$$

Thus, V_T also refers to the vector difference of the geostrophic wind at two successive pressure levels p_1 and p_2 ($p_2 < p_1$). In component form V_T reads

$$V_T = \left(-\frac{R}{f}\left[\frac{\partial \bar{T}}{\partial y}\right]_p \ln\left(\frac{p_1}{p_2}\right), \frac{R}{f}\left[\frac{\partial \bar{T}}{\partial x}\right]_p \ln\left(\frac{p_1}{p_2}\right)\right)$$

$$\text{.....................................(7)}$$

Using the hypsometric equation, eq. (7) can we written as

$$V_T = \left(-\frac{1}{f}\left[\frac{\partial(\Phi_2-\Phi_1)}{\partial y}\right], \frac{1}{f}\left[\frac{\partial(\Phi_2-\Phi_1)}{\partial x}\right]\right) = \frac{1}{f}k\times\nabla(\Phi_2-\Phi_1)$$

$$\text{...................(8)}$$

Thus, if the $V_g(p_1)$ is known at 850 hPa, then from the horizontal variation of mean temperature $\bar{T}$ of the layer bounded by pressure surfaces 850 hPa and 500 hPa, it is possible to find $V_g(p_2)$ at 500 hPa using (7). Define $\delta\Phi = \Phi_2 - \Phi_1$ then in the Figure, $\delta\Phi(x_2) > \delta\Phi(x_1)$ because $\delta z_2 > \delta z_1$; this implies that $v_T > 0$. Therefore at location x_2, the geostrophic wind will increase at pressure level $p_2 = p_1 - \delta p$.

Since geostrophic wind blows parallel to isobars with high pressure on the right while facing downstream in the northern hemisphere, therefore, isobars and isotherms are not parallel but make an angle to each other in a horizontal plane. This relationship is illustrated in Fiure where strengthening and turning of the geostrophic wind with height may be noted with cold air advection. Therefore, thermal wind relation finds its immediate use in correcting the data at different levels in the atmosphere and also in the ocean. However, the key question is: when will the wind rotate clockwise or counter clockwise with height? The knowledge of thermal wind helps in addressing this question.

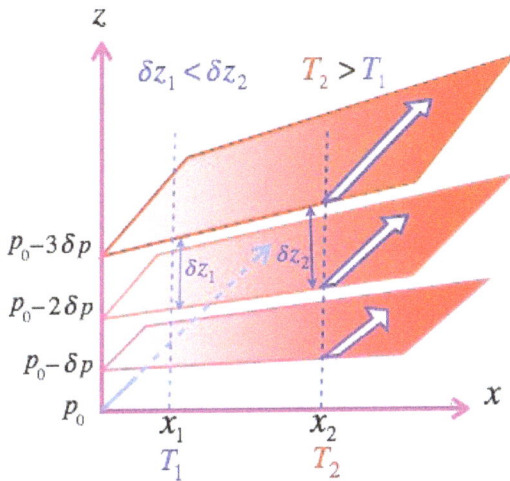

Relationship between vertical shear of the geostrophic wind and horizontal temperature gradient; Note that $\delta p > 0$, and pressure decreases with height. Also note that thickness of isobaric layer δz_1 at point x_1 is less than δz_2 at point x_2; that is, the thickness of isobaric layer increases with height on the east (or right) relative to the direction of surface wind (directed along y axis or northward) while one is positioned at the x_1 location.

The concept to thermal wind has also found application in explaining the warm core and cold core eddies in the ocean and relating the horizontal gradients of density to vertical variation of the geostrophic current. For the northern hemisphere, the rule – *Light on the Right* – holds in oceanography, which can be derived from geostrophic balance assuming the ocean in hydrostatic balance. Further, it can be noted from Figure that the *level of the free surface is higher in warm water* than that in the cold ocean water. The interpretation of the maxim *"Light on the Right"* is that if one stands in the direction of the current looking downstream in the northern hemisphere, then low density or light (warm) waters are on ones right hand side. To derive the thermal wind relation in the ocean, instead of isothermals (atmospheric case), the distribution of isopycnals showing the arrangement of water masses is used.

Now we examine, what happens to the geostrophic wind with height in the atmosphere if there is cold (warm) air advection into a warm (cold) area. This is illustrated in Figure, which shows the direction of the thermal wind in the disposition of isothermals with warm air on the right and cold on the left; and the geostrophic wind is directed from a colder region towards warm area. A geostrophic wind that turns counterclockwise with height (backing of the wind) is associated with cold air advection. In contrast to this, if the geostrophic wind turns clockwise with height (veering of the wind), it implies warm air advection by the geostrophic wind in the layer as illustrated in Figure b.

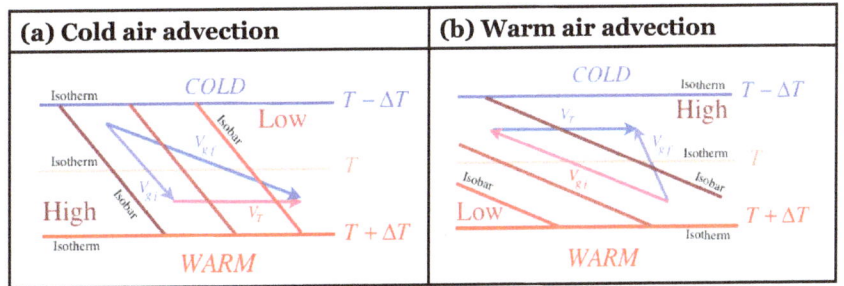

Relationship between turning of the wind and temperature advection: (a) backing of wind with height; (b) veering of the wind with height. Note the increase
in wind speed with height in case (a) as a result of cold air advection, but the
geostrophic wind speed decreases with height when there is warm air advection
as seen in case (b). The isobars are inclined to the isotherms; geostrophic wind
flows parallel to isobars, and the thermal wind is parallel to isotherms.

The use of thermal wind in the ocean is illustrated in Figure. Let us assume that the geostrophic current in the ocean has only the meridional component of velocity directed northward. Also if $\rho=\rho(x)$
and $\partial\rho/\partial x<0$; then it would mean that light (low density) water is on the right relative to the direction of the current. The thermal wind relation in the ocean says: *In the northern (southern) hemisphere, if the isopycnal slope upward to the left (right) across a current while looking downstream,
then current speed decreases with depth vanishing at some level, often referred to as "the level of
zero velocity". If the isopycnals slope downward to the left (right), then current speed increases
with depth in the northern (southern) hemisphere.* The strengthening of the current velocity with
depth has been illustrated in Figure for the Southern Hemisphere. It may be noted that the thermal
wind will anyway become zero at some depth where isopycnals become horizontal.

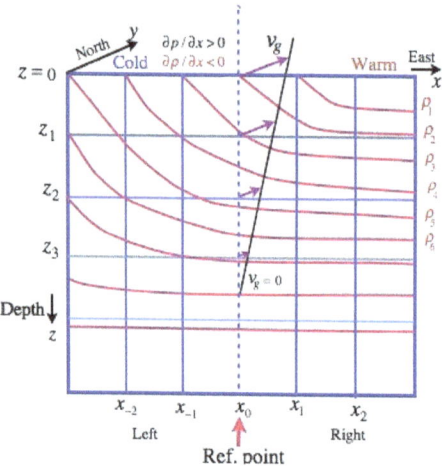

Illustration of thermal wind in the ocean. The direction of the geostrophic current is northward and the isopycnals
slope upward on the left while looking downstream to the current at a reference point x_o along the east. The density
distribution shows light waters on the right to the direction of current. Note that current velocity
decreases with depth as the thermal wind $v_T<0$. Accordingly the highest current velocity is at
the surface, which vanishes at some depth in the ocean.

The density of seawater depends both on salinity and temperature. Though salinity varies between
33 - 37 psu but temperature, expressed in degrees Celsius (°C) , shows large variation from warm
to subzero temperatures. For this range of salinity, seawater density would vary in the range 1022
- 1028 kg m⁻³. Hence, instead of density as a variable in the thermal wind relation, it is convenient
to use density anomaly (σ_t) which is express numerically as

$$\sigma(T,s,p)=\rho-\rho_0 ; \quad \rho_0=1000\ kg\ m^{-3}$$

....(9)

In terms of σ_t, the thermal wind relation takes the following form

$$\frac{\partial v_g}{\partial z}=-\frac{g}{f\rho_0}\frac{\partial\sigma_t}{\partial x}\qquad (z\ increasing\ upwards)$$

....(10a)

$$\frac{\partial u_g}{\partial z} = \frac{g}{f \rho_0} \frac{\partial \sigma_t}{\partial y}$$

....(10b)

Note that the reference density ρ_0 in (4) and (10) is the same. In Figure the disposition of isopycnal is given for the southern hemisphere (SH). In this figure, isopycnals slope downward to the left with respect to the direction of current velocity ($v_g > 0$) at the reference point. It can be inferred from that the geostrophic current speed in the ocean would decrease downward from the free surface because the thermal wind ($v_T < 0$) ; this means that disposition of isopycnals in the SH as given in Figure (with origin at the surface), weakens the southerly current at the deeper levels. However, the potential density contour 26.5 kg m^{-3} slopes downward on the right to the direction of current at 60 m near the coast, therefore the southerly current would strengthen below 60 m depth near the coast in accordance to the thermal wind relation.

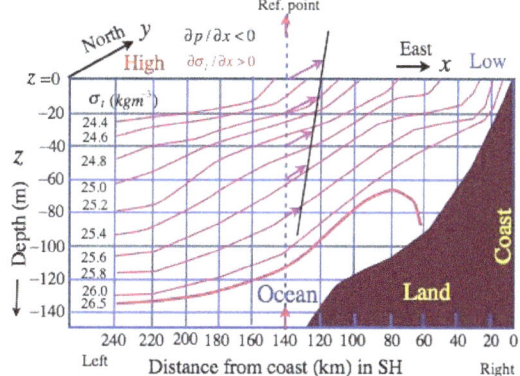

In SH, the geostrophic current is northward ($v_g > 0$), isopycnals slope downward on the left across the current shown at a reference point at 140 km, west to the coast. The density distribution shows light waters on the left with respect to the direction of current. Current velocity decreases with depth as the thermal wind ($v_T < 0$) for z decreasing downwards. Near the coast, the isopycnal slopes downward so its consequence can be imagined.

By applying the rule *"light on the right in northern hemisphere"*, it can be inferred that *a clockwise circulation* of seawater is around a *warm core eddy*; while *an anti-clockwise circulation* will be found around a *cold core eddy*. The circulation of these eddies changes due to stretching and shrinking of water columns. In the warm core eddy, the *lines of constant density will slope upward from the vortex centre on both sides*; whereas the lines of constant pressure will slope downward on both sides from the centre of the warm core eddy as shown in Figure (a). That is, the sea level height attains a maximum at the eddy centre. In case of a cold core eddy, the lines of constant pressure slope upwards as shown in Figure (b), while the lines of constant density slope downwards from the centre of the vortex.

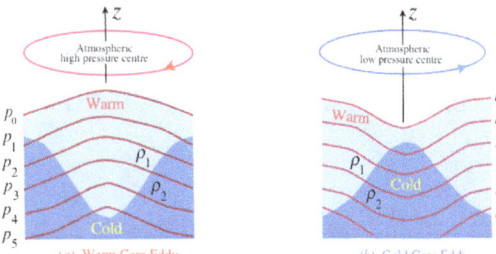

Sloping isobars in oceanic eddy circulation forced by atmospheric circulations.
(a) Warm core eddy: clockwise circulation with isobars sloping downwards and isopycnals upwards from the centre ;
(b) Cold core eddy: anti-clockwise circulation; isobars slope upward but isopycnals slope downwards from the centre.

Another important fact about ocean is that in the upper 1500 m depth the variation in the temperature are more prominent as compared to density; that is, the effect of vertical gradient $\delta_z S$ on salinity is much smaller than the effect of $\delta_z T$ on the temperature. Hence, increase or decrease of current velocity with depth can be assessed from a vertical section of temperature. Since the thermal wind relation in oceanography refers to the sloping isopycnals, therefore from a hydrographic section of any region, one can have an idea about the current speed perpendicular to a section (up to a depth of 1500 m) whether it is increasing or decreasing with depth based on the inclination of the thermocline in that oceanic region. Further, geostrophic currents in ocean can be related to the horizontal variations in the thickness of a water layer bounded by any two pressure levels.

Thus, the velocity of such currents one can estimate from the dynamic height (also *steric height*) of the ocean free surface. The steric effect, as discussed above, arises due to contraction and expansion of water columns due to T and S anomalies. When the *dynamic height* slopes upwards, the thermocline slopes downwards. This also implies that a given hydrostatic pressure is attained at lower depth in cold waters (say, near poles) than the warm waters of low latitudes. Hence, the dynamic height is a useful parameter in the study of ocean dynamics. The steric height (h) is introduced in oceanography as it related the changes in sea water density relative to that at standard temperature (0°C) and salinity (S=35 psu) ; mathematically, the steric height h can be derived from the hydrostatic balance that reads

$$h(z_1, z_2) = \frac{1}{g} \int_{p(z_1)}^{p(z_2)} \frac{dp}{\rho} \, ; \rho = \rho(T, S, p)$$

....(11)

There are two most important illustrations of the concept of thermal wind in ocean:

(i) western boundary current, *the Gulf Stream*, with slightly warmer water but slightly less salt than the average of the Sargasso Sea surface waters at the same latitude; and (ii) *zonal current* over the equator. From the hydrographic data and the thermal wind relation, it is possible to find the speed of the current at a given depth relative to another depth. In (i) the current speed is stronger on the west so isopycnals slope downwards which is associated with deepening of thermocline and rising steric height on the west of the boundary current. The current velocity in the Gulf Stream is greatest (ranging from 100 cm s⁻¹ to 300 cm s⁻¹) at the surface, which gradually decreases with depth reaching to 1 to 10 cm s⁻¹ between 1500 to 2000 m . The case (ii) can also be understood analogously.

The thermal wind relation has also found its application to explain the intensification of currents in the deep ocean. For example, the intensification of Antarctic Bottom Water (AABW) arises because the isothermal and isohalines at 4000 m in South Atlantic Ocean slope upward towards the South American coast; this is consistent with the northward thermal wind to increase in speed with depth.

Barotropic and Baroclinic Atmospheres

In an atmosphere, if the lines of constant pressure (isobars) are inclined to the lines of constant temperature (isotherms) or the lines of constant density (isopycnals), then such an atmosphere is called a baroclinic atmosphere. In a baroclinic ocean, isobars and isopycnals are inclined to each other. In a barotropic atmosphere or ocean, isobars and isopycnals are parallel to each other or coincide; that is, the vertical structure of the atmosphere or ocean is such that all the layers have same orientation vertically. However, if the isobars and the geopotential surfaces coincide in a barotropic fluid, then it is devoid of motion; but a baroclinic fluid is always in a state of motion.

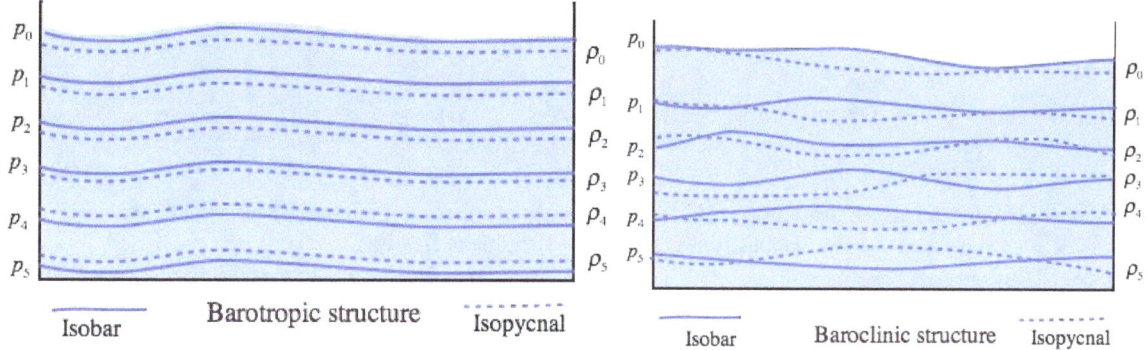

Barotropic and baroclinic structures of the stratified atmosphere and ocean.

It is worthwhile to illustrate the structure of barotropic and baroclinic states of the atmosphere or ocean. The barotropic structure of the ocean shows that density lines are parallel to the pressure lines (isobars); density is a function of pressure alone, $\rho=\rho(p)$; that is, the isobars and isopycnal surfaces do not intersect. In a barotropic atmosphere, the gradient of temperature on pressure surfaces (isobars) will identically vanish. *This means that if thermal wind vanishes, the atmosphere is barotropic and the geostrophic wind is independent of height.* The structure of a baroclinic ocean is also shown in Figure (right panel). The isobaric surfaces and isopycnal surface intersect in a baroclinic ocean/atmosphere because density depends on both temperature and pressure. That is, the density $\rho=\rho(p, T)$ is a function of both pressure and temperature. For understanding the meteorology of everyday weather events, analysis of states of a baroclinic atmosphere is of primary importance, for thermodynamic diagrams are used. The India Meteorological Department (IMD) uses the tephigram for this purpose, which will be introduced later in this course. However, for developing numerical models of weather prediction, much of the understanding can be achieved with the discrete analogues of the governing set of differential equations representing the barotropic atmosphere.

Vertical Motion in the Atmosphere

In the atmosphere, rising and sinking motions of air masses are as important as their horizontal motions. Indeed, the large-scale vertical motion is an order smaller than the horizontal velocity. As a consequence, the development of weather is rather slow, and it is generally said that weather changes evolve in 3-6 hours, therefore atmospheric motions are synoptic motions. Their horizontal scale of variation is more than 100 km. But strong vertical rising and sinking air motions are observed in convective developments leading to cloud and heavy rains or the so called high impact weather events that occur on a much smaller scales than that mentioned above. As a result of vertical motion in the atmosphere, surface pressure changes are noticeable that are related to the horizontal divergence ($\nabla.V$) as illustrated in Figure below.

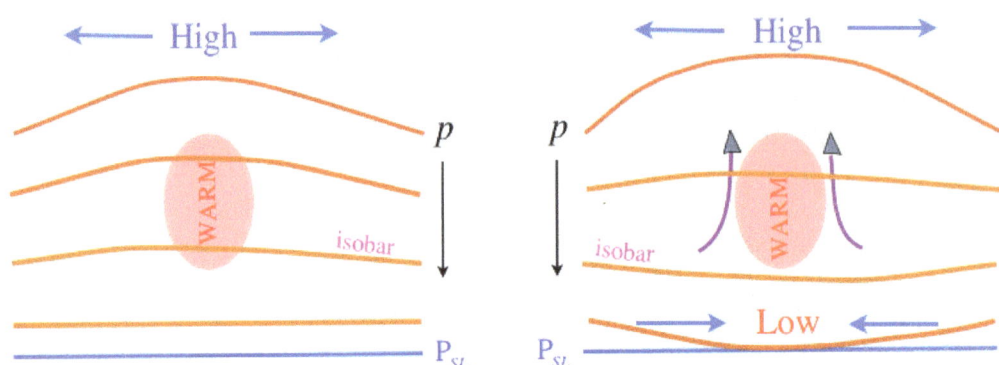

Adjustment of surface pressure to mid tropospheric heat source. P_{SL} indicates Sea Level Pressure. Isobars deflect up and down due to heating.

We suppose that a heat source generates local warming in the mid troposphere (~5 km height) as shown in Figure *(left panel)*. Because of the associated vertical movement of mid tropospheric air, the height of the upper-level pressure surfaces is raised above the warm anomaly; and as a consequence, there is a horizontal pressure gradient force at the upper levels driving a *divergent* upper level wind producing a low pressure system at the surface as shown in Figure (right panel). In other words, the net effect of the mid tropospheric heat source is to induce inflow that reduces surface pressure.

The wind circulation constantly distributes mass forming high and low pressure regions under the action of friction, flow curvatures induced by the Coriolis force, heating and cooling. Thus, there are two types of high pressure areas in the atmosphere; Figure clearly represents a upper thermal high. However, there is another class of upper high pressure centre that is called a dynamic high, which is, for instance, found over West Africa and many parts of the world. That feature invites inflow of air in to the high for it formation dynamically. Such dynamic high pressure areas are formed as result of redistribution of mass by winds in the upper levels which are intimately linked to the systems forming at the surface and vice-versa. The redistribution of mass accompanies, for example, the cold or warm air advection and the consequences of such an advection can be easily analyzed from the concept of thermal wind. The impact of upper level divergence forced dynamically or due to warm anomalies is to reduce further the initial surface pressure generating a surface low just below it. The horizontal pressure gradient associated with surface low will drive low-level convergence. The continuous fall or rise of the surface pressure is mainly determined by the degree of compensation of mass occurring between upper-level divergence and low-level convergence. *It may therefore be concluded that upper-level divergence forced dynamically or by mid tropospheric heat anomalies; and the associated low-level convergence is one of the most important parameters that induce weather changes.*

However, the key question is: *What is the relative importance of upper level divergence due to an upper high and low level convergence due a low pressure in producing weather in mid latitudes and tropical atmosphere?* This question is relevant in order to understand the cyclogenesis in mid latitudes and the tropics. If the atmosphere is stratified like in the midlatitudes, then the upper level anticyclonic circulation anomalies will be very efficient to cause cyclogenesis in the area. Where as, if the atmospheric column is warm and calm like that in the tropics, the low level convergence will be very effective in producing cyclogenesis conditions in the area. To summarize, the upper level anticyclonic anomalies in the midlatitudes and the low level convergence in a neutrally buoyant tropical atmosphere are crucial weather producing mechanisms.

We now look at other mechanisms, which would produce vertical motions in the atmospheric flows. Earth is a rotating planet and therefore the vortex lines are closed circles straddling around the planet's surface. However, from equator to pole, the radius of the latitude circles decrease and its value is zero at the poles. The shifting of air parcels from one latitude circle to another has important consequences for atmospheric motion as it may lead to stronger cyclonic circulations producing intense rising motions one generally notices in a cyclone. The mechanism is related to parcels changing their latitudinal positions. Similarly, if the fluid columns are stretched or compressed, it will result in change of the vertical component of the relative vorticity. Both these mechanisms are discussed below in the next section.

Change of Relative Vorticity

The fluids at rest rotate with the earth, but the fluid masses having spinning motion relative to the vertical axis possess *relative vorticity* (ζ) given in Cartesian coordinates as

$$\zeta = \frac{\partial v}{\partial x} - \frac{\partial u}{\partial y}$$ ....(12)

The *absolute vorticity* ($\zeta+f$) of a fluid in motion is the sum of relative vorticity (ζ) of the fluid column and the planetary vorticity (f). From the equations of motion with rotation, the vorticity equation can be derived by simple calculus as

$$\frac{d(\zeta+f)}{dt} + (\zeta+f)\nabla_h.V = 0$$ ....(13)

In eq. (13), $\nabla_h.V = \frac{\partial u}{\partial x} + \frac{\partial v}{\partial y}$ is the horizontal divergence. On substituting $\nabla_h.V = -\frac{\partial w}{\partial z} = -\frac{1}{h}\frac{dh}{dt}$ in (13),

with h as the height of the fluid column; then after some manipulations, one obtains an equation for the absolute potential vorticity that was first derived by Rossby (1940)

$$\frac{d}{dt}\left(\frac{\zeta+f}{h}\right) = 0$$ ....(14)

That is $\left(\frac{\zeta+f}{h}\right) = const.$; or in other words, the absolute potential vorticity is *conserved* along fluid

trajectories. The eq. (14) laid the foundation of modern meteorology and it is also referred to as the *guiding principal* of short-range numerical weather prediction. One can also derive the conservation of absolute potential vorticity form the *Bjerknes circulation theorem* for rotating fluids (Holton 2004, Dutton 1976), which is the generalization of the *Kelvin's circulation theorem* for non-rotating fluids. Since circulation is a measure of rotation around an area of a fluid column, vorticity can thus be defined as the ratio of *circulation* and *area* in the limit of area tending to zero.

There are two important mechanisms by which the relative vorticity could change in the atmosphere and ocean:

1. Vertical stretching or shrinking of fluid columns.

2. Change of latitude by air mass or water mass columns of constant height.

The *first mechanism of change of relative vorticity* in the oceans and atmosphere has been demonstrated in Figure. The p*ositive relative vorticity* is produced in regions of low-level convergence while *negative relative vorticity* is produced in the locations of low level divergence. Low-level convergence could be produced due to heating in the column and the fall in surface pressure would be accompanied with the rising motion. When the fluid column is stretched due to vertical motion, it will, in turn, strengthen inflow at the surface. This will be accompanied with (i) enhanced cyclonic rotations of the fluid column due to Coriolis force, (ii) increase in its height and (iii) reduced diameter. On the contrary, if the fluid column shrinks, there will be outflow at the surface which will give rise to anticyclonic rotation under the action of the Coriolis force and larger vortex diameter. The continued compression of the fluid column will then strengthen the low level divergence. *Thus stretching and compression of the vortex would produce vertical motions, which in turn change the relative vorticity.*

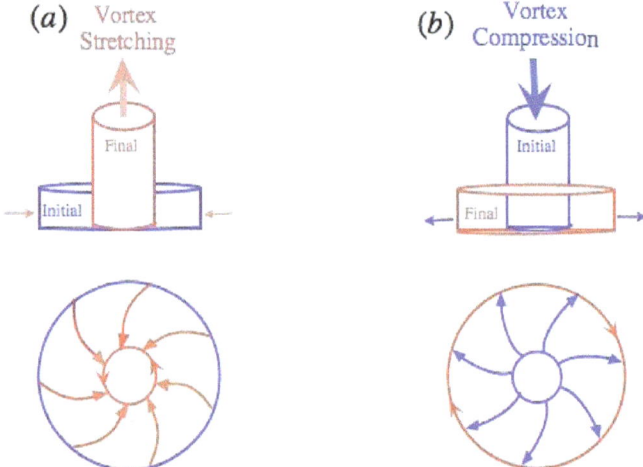

Production of relative vorticity by stretching and shrinking of air columns.
(a) Vertical stretching requires horizontal inflow, which produces cyclonic rotations.
(b) Vertical shrinking requires horizontal outflow, which produces anticyclonic rotations.

The *second mechanism of relative vorticity change* explains the development of positive and negative vorticity in fluid columns when they move from one latitude to another. The mechanism is founded upon the conservation of absolute potential vorticity of the parcels or fluid columns. For a column of constant height, the absolute vorticity is conserved (sum of planetary vorticity and relative vorticity). Assume that a column placed on 30°N has no rotational motion ($\zeta = 0$) initially. If it moves either way along this latitude then its absolute vorticity will not change, and its relative vorticity will continue to remain zero. However, if a fluid column on the equator with ($\zeta = 0$), moves northward or southward, then negative (anticyclonic) vorticity will be generated in order to conserve the absolute vorticity. Suppose a column with ($\zeta = 0$) from a higher latitude moves equatorward, then cyclonic (positive) vorticity will be produced as the column has shifted towards latitudes with smaller Coriolis parameter f. If the earth's topography is included in the discussion, then conservation of total potential vorticity (absolute vorticity divided by height of the topography) is the necessary condition.

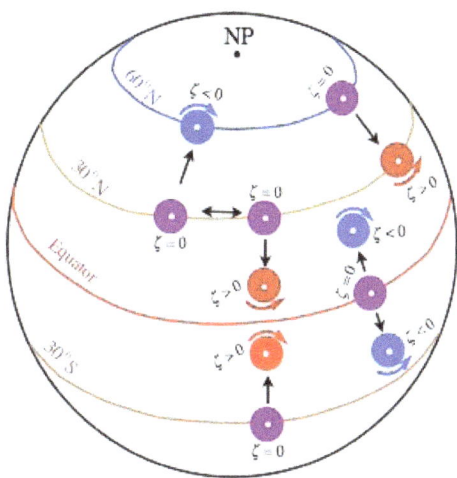

Absolute vorticity (planetary vorticity + relative vorticity) is conserved as fluid columns of constant height change latitude during their movement in the oceans and atmosphere.

The conservation of potential vorticity is therefore the guiding principle of short-range weather forecasting (1-3 days). The conservation of potential vorticity is relevant for the dynamics of both the atmosphere and oceans. The key point to be noted here is that the variation of Coriolis force in the meridional direction is a significant source of cyclonic or anticyclonic vorticity. This is referred to as *the beta effect* in the physics of the ocean and atmosphere. In a barotropic atmosphere, due to the beta effect (i.e. meridional gradient of planetary vorticity), *Rossby waves* are produced which are meteorological important waves propagating upstream in the westerly flow (mean zonal velocity $\bar{u}$) with the zonal phase speed (v), which can be derived from (14) as

$$v = \bar{u} - \frac{\beta}{k^2 + l^2}, \; k,l \text{ are wavenumbers along } x \text{ and } y \text{ axes}$$

....(15)

Weather Map

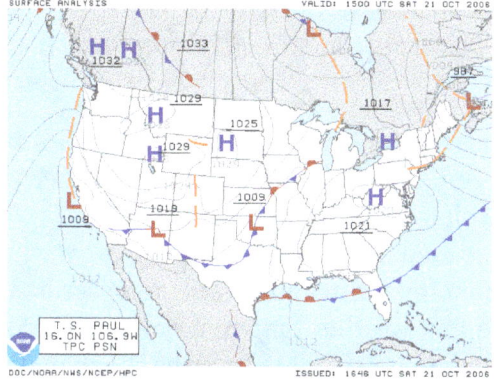

A surface weather analysis for the United States on October 21, 2006.

A weather map displays various Meteorological features across a particular area at a particular point in time and has various symbols which all have specific meanings. Such maps have been in use since the mid-19th century and are used for research and weather forecasting purposes. Maps using isotherms show temperature gradients, which can help locate weather fronts. Isotach maps, analyzing lines of equal wind speed, on a constant pressure surface of 300 mb or 250 mb show where the jet stream is located. Use of constant pressure charts at the 700 and 500 hPa level can indicate tropical cyclone motion. Two-dimensional streamlines based on wind speeds at various levels show areas of convergence and divergence in the wind field, which are helpful in determining the location of features within the wind pattern. A popular type of surface weather map is the surface weather analysis, which plots isobars to depict areas of high pressure and low pressure. Cloud codes are translated into symbols and plotted on these maps along with other meteorological data that are included in synoptic reports sent by professionally trained observers.

History

Sir Francis Galton, the inventor of the weather map.

The use of weather charts in a modern sense began in the middle portion of the 19th century in order to devise a theory on storm systems. During the Crimean War a storm devastated the French fleet at Balaklava, and the French scientist Urbain Le Verrier was able to show that if a chronological map of the storm had been issued, the path it would take could have been predicted and avoided by the fleet.

In England, the scientist Francis Galton heard of this work, as well as the pioneering weather forecasts of Robert Fitzroy. After gathering information from weather stations across the country for the month of October 1861, he plotted the data on a map using his own system of symbols, thereby creating the world's first weather map. He used his map to prove that air circulated clockwise around areas of high pressure; he coined the term 'anticyclone' to describe the phenomenon. He was also instrumental in publishing the first weather map in a newspaper, for which he modified the pantograph (an instrument for copying drawings) to inscribe the map onto printing blocks. *The Times* began printing weather maps using these methods with data from the Meteorological Office.

The introduction of country-wide weather maps required the existence of national telegraph networks so that data from across the country could be gathered in real time and remain relevant for subsequent analysis. The first such use of the telegraph for gathering data on the weather was the *Manchester Examiner* newspaper in 1847:

> ...led us to inquire if the electric telegraph was yet extended far enough from Manchester to obtain information from the eastern counties...inquiries were made at the following places; and answers were returned, which we append...

It was also important for time to be standardized across time zones so that the information on the map should accurately represent the weather at a given time. A standardized time system was first used to coordinate the British railway network in 1847, with the inauguration of Greenwich Mean Time.

In the USA, The Smithsonian Institution developed its network of observers over much of the central and eastern United States between the 1840s and 1860s once Joseph Henry took the helm. The U.S. Army Signal Corps inherited this network between 1870 and 1874 by an act of Congress, and expanded it to the west coast soon afterwards. At first, not all the data on the map was used due to a lack of time standardization. The United States fully adopted time zones in 1905, when Detroit finally established standard time.

20th Century

Light tables were important to the construction of surface weather analyses into the 1990s

The use of frontal zones on weather maps began in the 1910s in Norway. Polar front theory is attributed to Jacob Bjerknes, derived from a coastal network of observation sites in Norway during World War I. This theory proposed that the main inflow into a cyclone was concentrated along two lines of convergence, one ahead of the low and another trailing behind the low. The convergence line ahead of the low became known as either the steering line or the warm front. The trailing convergence zone was referred to as the squall line or cold front. Areas of clouds and rainfall appeared to be focused along these convergence zones. The concept of frontal zones led to the concept of air masses. The nature of the three-dimensional structure of the cyclone would wait for the development of the upper air network during the 1940s. Since the leading edge of air mass changes bore resemblance to the military fronts of World War I, the term "front" came into use to represent these lines. The United States began to formally analyze fronts on surface analyses in late 1942, when the WBAN Analysis Center opened in downtown Washington, D.C.

In addition to surface weather maps, weather agencies began to generate constant pressure charts. In 1948, the United States began the Daily Weather Map series, which at first analyzed the 700 hPa level, which is around 3,000 metres (9,800 ft) above sea level. By May 14, 1954, the 500 hPa surface was being analyzed, which is about 5,520 metres (18,110 ft) above sea level. The effort to automate map plotting began in the United States in 1969, with the process complete in the 1970s. Hong Kong completed their process of automated surface plotting by 1987.

By 1999, computer systems and software had finally become sophisticated enough to allow for the ability to underlay on the same workstation satellite imagery, radar imagery, and model-derived fields such as atmospheric thickness and frontogenesis in combination with surface observations to make for the best possible surface analysis. In the United States, this development was achieved when Intergraph workstations were replaced by n-AWIPS workstations. By 2001, the various surface analyses done within the National Weather Service were combined into the Unified Surface Analysis, which is issued every six hours and combines the analyses of four different centers. Recent advances in both the fields of meteorology and geographic information systems have made it possible to devise finely tailored products that take us from the traditional weather map into an entirely new realm. Weather information can quickly be matched to relevant geographical detail. For instance, icing conditions can be mapped onto the road network. This will likely continue to lead to changes in the way surface analyses are created and displayed over the next several years.

Plotting of Data

Present weather symbols used on weather maps

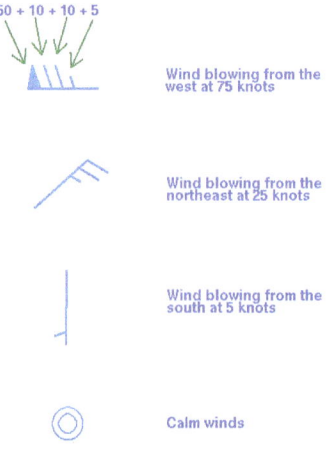

Wind barb interpretation

A station model is a symbolic illustration showing the weather occurring at a given reporting station. Meteorologists created the station model to plot a number of weather elements in a small space on weather maps. Maps filled with dense station-model plots can be difficult to read, but they allow meteorologists, pilots, and mariners to see important weather patterns. A computer draws a station model for each observation location. The station model is primarily used on surface-weather maps, but can also be used to show the weather aloft. A completed station-model map allows users to analyze patterns in air pressure, temperature, wind, cloud cover, and precipitation.

Station model plots use an internationally accepted coding convention that has changed little since August 1, 1941. Elements in the plot show the key weather elements, including temperature, dew-point, wind, cloud cover, air pressure, pressure tendency, and precipitation. Winds have a standard notation when plotted on weather maps. More than a century ago, winds were plotted as arrows, with feathers on just one side depicting five knots of wind, while feathers on both sides depicted 10 knots (19 km/h) of wind. The notation changed to that of half of an arrow, with half of a wind barb indicating five knots, a full barb ten knots, and a pennant flag fifty knots.

Because of the structure of the SYNOP code, a maximum of three cloud symbols can be plotted for each reporting station that appears on the weather map. All cloud types are coded and transmitted by trained observers then plotted on maps as low, middle, or high-étage using special symbols for each major cloud type. Any cloud type with significant vertical extent that can occupy more than one étage is coded as low (cumulus and cumulonimbus) or middle (nimbostratus) depending on the altitude level or étage where it normally initially forms aside from any vertical growth that takes place. The symbol used on the map for each of these étages at a particular observation time is for the genus, species, variety, mutation, or cloud motion that is considered most important according to criteria set out by the World Meteorological Organization (WMO). If these elements for any étage at the time of observation are deemed to be of equal importance, then the type which is predominant in amount is coded by the observer and plotted on the weather map using the appropriate symbol. Special weather maps in aviation show areas of icing and turbulence.

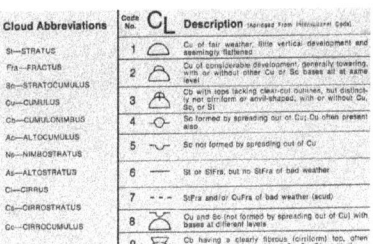

Low étage (Sc,St) and upward-growing vertical (Cu, Cb)

No.	C_M	Description (Abridged From International Code)
1		Thin As (most of cloud layer semitransparent)
2		Thick As, greater part sufficiently dense to hide sun (or moon), or Ns
3		Thin Ac, mostly semitransparent; cloud elements not changing much and at a single level
4		Thin Ac in patches; cloud elements continually changing and/or occurring at more than one level
5		Thin Ac in bands or in a layer gradually spreading over sky and usually thickening as a whole
6		Ac formed by the spreading out of Cu or Cb
7		Double-layered Ac, or a thick layer of Ac, not increasing; or Ac with As and/or Ns
8		Ac in the form of Cu-shaped tufts or Ac with turrets
9		Ac of a chaotic sky, usually at different levels; patches of dense Ci are usually present also

Middle étage (Ac,As) and downward-growing vertical (Ns)

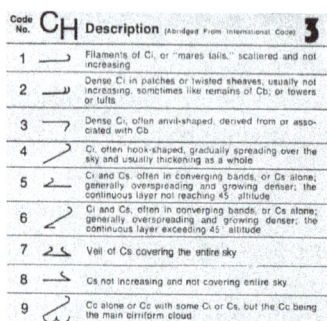

High étage (Ci,Cc,Cs)

Types

Aviation Maps

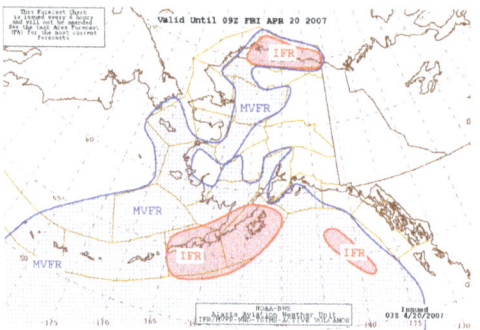

Alaskan aviation weather map

Aviation interests have their own set of weather maps. One type of map shows where VFR (visual flight rules) are in effect and where IFR (instrument flight rules) are in effect. Weather depiction plots show ceiling height (level where at least half the sky is covered with clouds) in hundreds of feet, present weather, and cloud cover. Icing maps depict areas where icing can be a hazard for flying. Aviation-related maps also show areas of turbulence.

Constant Pressure Charts

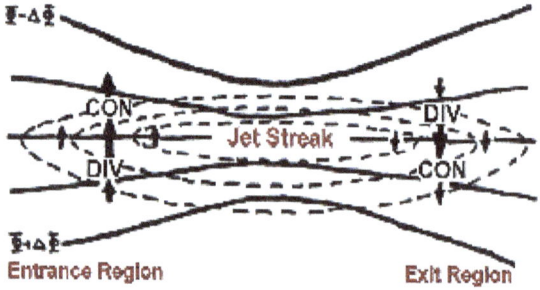

An upper level jet streak. DIV areas are regions of divergence aloft, which usually leads to surface convergence and cyclogenesis.

Constant pressure charts normally contain plotted values of temperature, humidity, wind, and the vertical height above sea level of the pressure surface. They have a variety of uses. In the mountainous terrain of the western United States and Mexican Plateau, the 850 hPa pressure surface can be a more realistic depiction of the weather pattern than a standard surface analysis. Using the 850 and 700 hPa pressure surfaces, one can determine when and where warm advection (coincident with upward vertical motion) and cold advection (coincident with downward vertical motion) is occurring within the lower portions of the troposphere. Areas with small dewpoint depressions and are below freezing indicate the presence of icing conditions for aircraft. The 500 hPa pressure surface can be used as a rough guide for the motion of many tropical cyclones. Shallower tropical cyclones, which have experienced vertical wind shear, tend to be steered by winds at the 700 hPa level.

Use of the 300 and 200 hPa constant pressure charts can indicate the strength of systems in the lower troposphere, as stronger systems near the Earth's surface are reflected as stronger features at these levels of the atmosphere. Isotachs are drawn at these levels, which a lines of equal wind speed. They are helpful in finding maxima and minima in the wind pattern. Minima in the wind pattern aloft are favorable for tropical cyclogenesis. Maxima in the wind pattern at various levels of the atmosphere show locations of jet streams. Areas colder than −40 °C (−40 °F) indicate a lack of significant icing, as long as there is no active thunderstorm activity.

Surface Weather Analysis

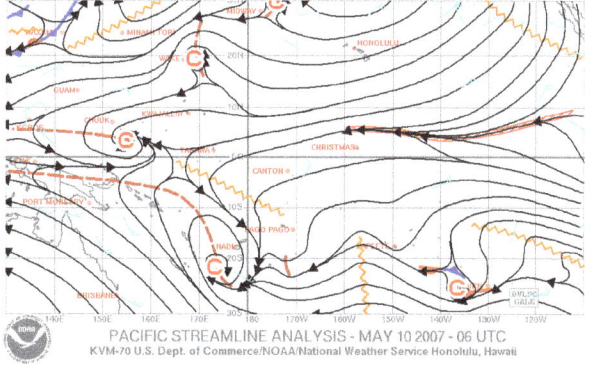

Streamline analysis of the tropical Pacific Ocean

A surface weather analysis is a type of weather map that depicts positions for high and low-pressure areas, as well as various types of synoptic scale systems such as frontal zones. Isotherms can be drawn on these maps, which are lines of equal temperature. Isotherms are drawn normally as solid lines at a preferred temperature interval. They show temperature gradients, which can be useful in finding fronts, which are on the warm side of large temperature gradients. By plotting the freezing line, isotherms can be useful in determination of precipitation type. Mesoscale boundaries such as tropical cyclones, outflow boundaries and squall lines also are analyzed on surface weather analyses.

Isobaric analysis is performed on these maps, which involves the construction of lines of equal mean sea level pressure. The innermost closed lines indicate the positions of relative maxima and minima in the pressure field. The minima are called low-pressure areas while the maxima are called high-pressure areas. Highs are often shown as H's whereas lows are shown as L's. Elongated areas of low pressure, or troughs, are sometimes plotted as thick, brown dashed lines down the trough axis. Isobars are commonly used to place surface boundaries from the horse latitudes poleward, while streamline analyses are used in the tropics. A streamline analysis is a series of arrows oriented parallel to wind, showing wind motion within a certain geographic area. C's depict cyclonic flow or likely areas of low pressure, while A's depict anticyclonic flow or likely positions of high-pressure areas. An area of confluent streamlines shows the location of shearlines within the tropics and subtropics.

The information collected by observers with radiosondes or manually is generally broadcast to all meteorological centres of different countries on the WMO network. This information is interpreted and depicted on global or regional charts for interpretation and visualization of evolving weather. Since meteorological waves are long waves, their evolution is slow and movement is along characteristic lines (or the axes). It is thus possible to predict weather with the help of synoptic charts by following the persisting direction suggested by the axes along which movement of weather systems such as low (troughs) or high (ridges) pressure systems would happen. The pressure systems of larger expanse are the so-called "centres of action" as referred by C.G. Rossby (1939) in a paper on "Planetary flow patterns in the atmosphere". The cold and warm air advection into an area could also be studied from the synoptic charts as much of the weather evolves due to advective processes. The mean state of the circulation at mandatory levels (as defined by the World Meteorological Organization) could be studied with mean global synoptic charts, which display the positions of the centres of actions pertaining to intraseasonal and interannual variations of dominant phenomena, such as the PNA, monsoons or the El Niño among many other features of the circulation.

The Monsoon Circulation

The story of the meteorology of Indian monsoons begins with the early essay of E. Halley (1676) - famous for discovering Halley's comet, and a better explanation later by G. Hadley (1735) of trade winds in the tropics. From these two great expositions also originates the basic understanding of processes that explains the general circulation of the atmosphere with trades in the tropics and the westerlies in midlatitudes. The broader Asian Summer Monsoon (ASM) includes the southwest summer monsoon over India, Mei-yu in China, Bei-yu in Japan and the Indonesian-Australian monsoon. Thus, its geographical region is spread over the Indian Ocean, India, Australia, Western Pacific, Indonesia, East China, and Japan. For pedagogical purpose, we discuss the mean fields of the Indian monsoon circulation over the region that includes the Indian Ocean, the Indian peninsula, the East-African Highlands, and deserts of Arabia. That is, the region on the globe bounded latitudinally by 25°S and 45°N, and longitudinally by 30°E and 125°E.

The monsoon circulation is forced by the thermal contrast of land and ocean arising due to large differences in their specific heat capacities. It is also important to emphasize that the notion of differential heating between land and ocean is central to the establishment of summer monsoon over India. However, once the monsoon sets in, the latent heat released in deep convective towers assumes the key role in maintaining the circulation. Further, one can define monsoons in several ways. This is however done towards the end of the section series when we present a fairly comprehensive and complete exposition with the help of mathematical models of heat-induced tropical circulations based on the work of Adrian Gill (1982). In the equatorial region, the most important zonally oriented Walker circulation is also forced by heating of the atmospheric column by the release of latent heat in clouds. The Gill's model successfully explains both the zonally and meridionally oriented heat-induced circulation.

We now examine the mean circulation (i.e. steady winds) during July at 850 hPa (about 1.5 km height above surface) in the monsoon region. The direction of the airflow is depicted with contour lines with an arrow (streamlines) in the figure. The lines spiral inwardly in some locations to form closed circular regions of low pressure (marked C) relative to surroundings; whereas in some locations the lines spin outwardly from closed circular regions (marked A) of higher pressure relative to surroundings. The regions marked L represent convergence zones (cyclonic circulation of air) and those marked A represent divergence zones (anticyclonic circulation of air). Both these centres of action produce distinct weather states: cloudy and rainy conditions over regions marked L and fair weather and clear conditions over the regions marked A. The synoptic charts display more information such as the distribution of surface pressures, temperatures, winds, rains and cloud cover in octas. In Figure, the low-pressure centre over northwest India and Pakistan is very dominant, while some of the isobars first deflect northward over the Arabian Sea and then curve southwards over the Bay of Bengal. This is very familiar depiction of pressure distribution during July.

Mean Wind Flow: July 850 hPa (YP Rao 1976, Southwest Monsoon, IMD Publication)

A clear observation may be made from Figure that isobars are narrowly packed over the Arabian Sea where winds up to 30 knots are found; but notice the spreading of isobars in the regions of weak winds. The surface is both geostrophic and ageostrophic depending on the region. As one moves along an isobar from west to east, it is noted that the *wind barbs* make an angle with the local direction of the isobars as seen over land areas (ageostrophic flow), but in some areas their direction is parallel to the isobars especially over the oceanic regions (geostrophic flow). Land areas are *rough* whereas the oceanic regions are *smooth*. Hence wind experiences greater friction over land causing wind barbs to deflect, as compared to the oceanic regions. In the meteorological parlance, *easterly* will mean *winds coming from east* for an observer standing with face towards east; similarly, *westerly* will mean *winds coming from west*. Analogously, the south-westerlies will refer to winds coming from the southwest direction etc.

The circulation of the air is plotted in Figure with *wind barbs* to represent both magnitude and direction of the wind at any location. Wind barbs *point in the direction in which wind is blowing*. A *full barb* represent a wind speed of 10 knots (~5 ms⁻¹, and *half barb* is used 5 knots wind speed. A *solid triangle* on the line representing the direction will mean a 50 knots wind.

We have southwesterly winds during the summer monsoon (also called south-west monsoon) over India when the country receives a major quantum of its annual rainfall. The distribution of surface mean pressure and the associated circulation during July in Figure shows that there is a well marked low pressure (994 hPa) system, also called the "heat low", over Rajasthan, Punjab and Pakistan region, which even extends to Iran sometimes; and a weak low (1000 hPa) on the east over Assam region. The isobars are convex (opening southward) due to their inclination southward over the peninsula and then become concave (opening towards north and northwest) over the Bay of Bengal. Almost all the isobars have similar structure in this region.

The explanation for the equator ward deflection of isobars over southern India was first attempted by S.K. Banerji (1923) considering the region's all dominant mountain ranges – the Deccan Plateau, Himalayas on the north and Khasi Hills of Burma (Myanmar) on the east – acting as orographic barriers to the monsoon current. Although S.K. Banerji correctly estimated the vertical velocity as a result of breaking of 850 hPa geostrophic wind differently over land (rough) and sea (smooth) at surface level, his mathematical analysis failed to explain the observed deflection as he did not take into account the stretching of the planetary vortex lines as a result of orography induced vertical motions (or the conservation of absolute potential vorticity). It has been explained earlier that when columns shift latitude, a change in the relative vorticity would be noted, which in turn modifies the circulation. During monsoon, the depth of the column upstream in the southwesterly current over Arabian Sea shrinks drastically due to the steep orography of the Deccan Plateau as the current hits the India's Malabar Coast. This will produce strong cyclonic (positive) vorticity in the current and consequently it will shift equator ward in order to conserve the absolute potential vorticity.

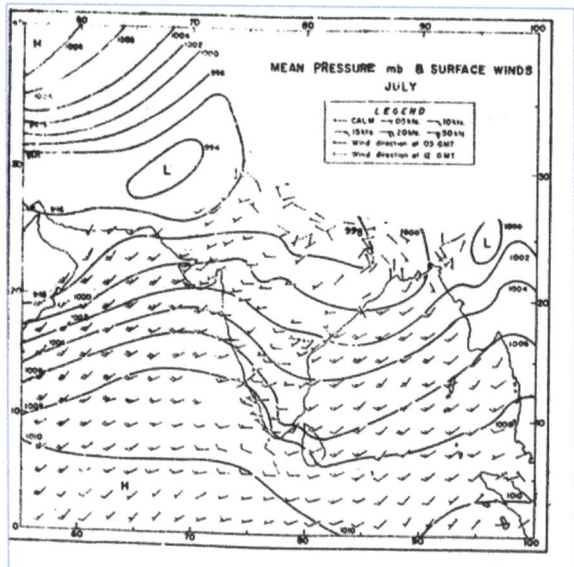

July Mean Pressure and Surface winds (YP Rao, Southwest Monsoon, IMD Publ. 1976)

The wind field in Figure shows that over the Arabian Sea in the region around 60°E in the latitudinal belt 10°-20°N, the wind speeds exceed 25 Knots (~12.5 m s^{-1}). This is the *Somali Jet* that brings moist marine air over the heated Indian subcontinent. The strength and time of establishment of the Somali Jet over the Arabian Sea has a direct bearing on the performance of the southwest monsoon over India. On a typical day Somali Jet speed could reach as high as 50 m s^{-1} (Krishnamurti

2010) at 850 hPa. The formation and disappearance of the Somali Jet is directly related to the heat sources and sinks in the monsoon region. Another key feature of the monsoon circulation is the existence of easterlies-south easterlies (though weak) along the foothills of Himalayas and westerlies-south westerlies on the south relative to the axis of the low pressure region, the so-called *Monsoon Trough* over the Indo-Gangetic Plain that extends from Pakistan to the Bay of Bengal. The monsoon trough is one of key determinants of the active-break-active cycle in the Indian summer monsoon (P.K. Das 1997). The oscillations of the monsoon trough are clearly visible in the sequence of daily 850 hPa geopotential charts, and its north-south oscillations directly affects the monsoon activity: if the monsoon trough moves north from its average position, it marks the *break in monsoon (cessation of rains)* over Central India and the Indo-Gangetic Plain; but, it will brings copious rains in the foothills of Himalayas, and would also enhances the monsoon activity over southern parts of India.

Highest rainfall amount are noted over the Western Ghats on the Deccan Plateau in Figure, while lowest rainfall amounts are recorded over western Rajasthan (the Thar Desert). There are also some regions where rainfall is low (e.g. Odissa, parts of Madhya Pradesh, Rayalaseema in Andhra Pradesh). The distribution of rainfall every year is different to the extent that the *interannual variations* in the rainfall are strong. The situation is such that in some year, there may be flooding during the monsoon period, but the same region could be drought-prone in the subsequent year. The *vagaries of monsoon* are well known to the people of the Indian subcontinent. That is, the *intra seasonal variations* directly determine the overall performance of the monsoon; and, the all-India rainfall is its key indicator which is being since the time of Sir Gilbert Walker (1923). If the all-India rainfall during a particular year is close to the long-term average precipitation of past monsoons, the monsoon is termed *normal* and if it exceeds the normal value, it is an above normal or a good monsoon year; however, when rainfall is less than the normal, it is below normal, but if the all India rainfall is below 20%, the country faces severe drought, while if the rainfall is 20% above the normal, there are devastating floods in several parts of the country.

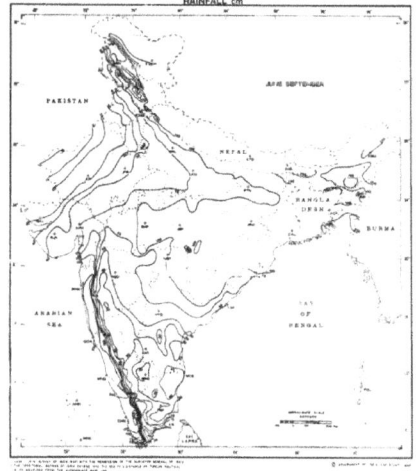

June-September Rainfall (YP Rao 1976, South west Monsoon, IMD Publication)

The rainfall is related to the weather systems; a low pressure system is rain bearing but high pressure systems will not support cloud formation and its subsequent growth. During the winter season, the winds reverse and the country experiences winter monsoon over Tamil Nadu and the adjoining coastal regions. However, the perturbations or the so-called *western disturbances* arriving from west bring cold air over northwestern India and cause widespread clouding accompanied with rain events in the north and northwest India. On the arrival of a western disturbance (moving like a front), the region experiences milder temperatures during winters. A very good account of the Indian monsoons has been given by P.K. Das (1997) in his book entitled "Monsoons", which also discusses the long-range forecasting of monsoon rainfall of Sir Gilbert Walker (1923) in detail.

Global Atmospheric Circulation

The planet earth is heated at the equator by incoming radiation from sun and cooled at the poles by outgoing infrared radiation. If the earth were a nonrotating planet, heating at the equator and cooling at the poles would produce only one cell with hot air rising over the equator, then moving pole ward aloft and finally sinking at the poles. The compensating surface circulation as shown in the Figure *a*, will be directed equator ward from high pressure at poles towards low pressure in tropics. That is, both in the upper levels and at surface, *flow is down the pressure gradient* like those studied in a fluid dynamics laboratory. The closed circulation in height-latitude plane thus produced is known as the *Hadley Cell*.

On a rotating planet, the Hadley cell is unstable and *breaks up into three distinct cells* that dominate the atmosphere (Figure *b*) in three different latitudinal belts. The *Hadley cell* is present in the equatorial region (thermally forced by heating at the equator), the *Ferrel cell* (indirect) in the middle latitudes and the *Polar cell* in the polar regions. The thermally forced circulations are fundamentally *direct cells* but the Ferrel cell is an *indirect cell*. An interesting direct cell forced by off-equator heating is the *"monsoon cell"* over India, which is forced by heating of Tibetan Plateau (the so-called elevated heat source) with rising motion over the Indian subcontinent and sinking motion over the Indian Ocean. The monsoon cell circulation is reverse of the Hadley cell circulation and it is therefore sometimes referred to as the *reverse Hadley cell*.

The general circulation of the atmosphere is depicted in Figure shows the Hadley cells tropics, Ferrel cells in middle latitudes and the Polar cells. In the lower branch of the Hadley cell, the deflected flow directed towards equator arises due to combined action of pressure gradient (surface pressure is lower in the equatorial region as compared to subtropical latitudes)and the Coriol is force. The easterlies (or the trade winds) therefore arise in order conserve the angular momentum of the earth. The lighter hot air rises in the ascending branch of the Hadley cell which moves pole ward aloft and the denser cold air sinks in the descending branch of the Hadley cell in the subtropical high pressure region. In the upper tropospheric branch of the closed Hadley cell, flow accelerates constantly as the potential energy of the parcels is converted into kinetic energy that increases the speed of parcels moving from equator to subtropical regions where air sinks. In this context, it is worthwhile to underscore the importance of tilting of the tropopause towards poles. Further, in the lower branch of the Hadley circulation there is significant cross-isobaric flow. The geopotential height of the isobars in the equatorial upper troposphere are greater than their corresponding height in the midlatitudes; hence the upper tropospheric pressure over the equatorial region is higher as compared to that in the midlatitude upper troposphere.

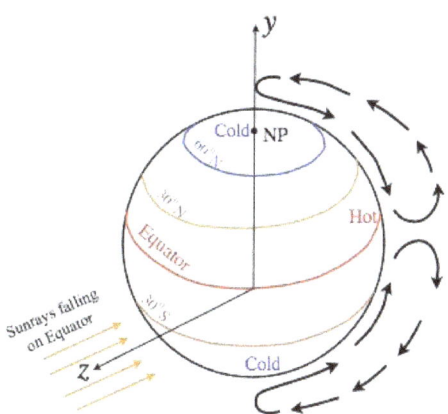

Movement of air on a nonrotating earth with sun directly above equator. Warm air rises and moves poleward aloft. The cold sinking air then moves equatorward. This type of air motion is known as the Hadley cell circulation.

The Ferrel cell is thermally indirect as the colder air rises over the region that separates the mid latitudes and polar latitudes; the cold air moves equatorward in the upper branch of the Ferrel cell and becomes relatively warmer which sinks in the subtropical high pressure region. Therefore, *Ferrel cell is thermally indirect because cold air rises and warm air sinks while Hadley cell is thermally direct because warm rises over the region of heating and cold air sinks farther away.* The existence of these cells is intimately liked to the rotation of the earth, heating over the equator and cooling in polar latitudes and the distribution of pressure and winds.

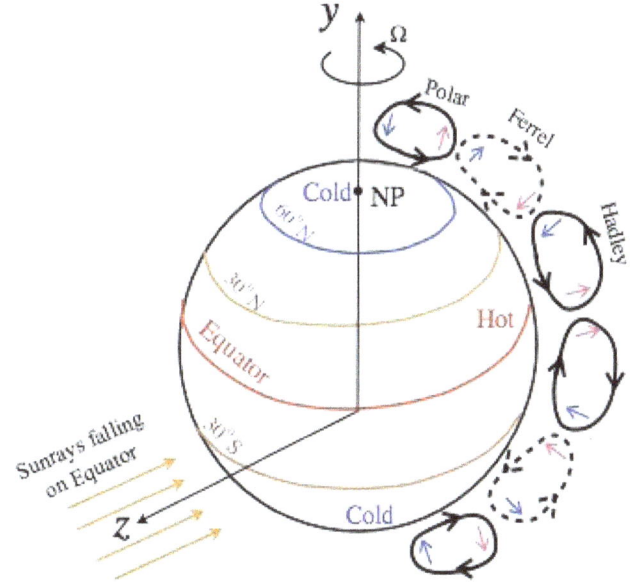

General circulation of the atmosphere: Idealized air circulation on a rotating earth with heating at equator and cooling at Poles. The arrows within a cell denote rising or descending cold or warm air. In the Hadley (direct) cell, warm air rises at the equator and cold air sinks. On the contrary, warm air descends and cold air rises in the Ferrel (indirect) cell.

Broadly, low pressure at the surface is topped by high pressure in the upper troposphere and vice versa. Consequently the wind flow is of opposite sign in the lower and upper tropospheric atmosphere with air velocity vanishing at some middle level in the troposphere. This is consistent with the thermal wind relation. Much of the information on the atmospheric circulation can be inferred from the surface pressure charts and the associated wind circulation. We, therefore, discuss only the January and July mean surface pressure charts, which are typical of the winter and summer seasons on the globe.

The horizontal distribution of surface pressures and of winds is shown in the following figures. The maps show regions of *high (H)* and *low (L)* pressures with the associated large-scale circulation. These highs and lows are dominant *centres of action* that are quasi-permanent features of the global atmospheric circulation. Variations in their strength and positions represent seasonal changes or the variability of the mean circulation during a particular month. A keen inspection of Figures is highly rewarding insofar as it concerns to understanding of the intraseasonal patterns that evolve on a monthly scale. The high pressure regions are located at around 30° latitude (first noted by Ferrel) in both the hemispheres. January is the representative month for winter while July for the summer in the Northern Hemisphere. During summer, the dominant variance of the atmospheric general circulation is monsoon circulation over India that arises due to excessive heating of the elevated landmass (the Tibetan Plateau) and its persistence which gives a quantum jump in the rainfall.

One may notice from Figures, that in the Southern Hemisphere the location of highs are *quasi-permanent* around 30°S, but *in the Northern Hemisphere the high pressure regions in January around 30°N are replaced by low pressure centres in July*. The key inference of this comparison is that there is an oscillation on the seasonal scale, which is gradual but sure. Why does northern hemisphere only witness such oscillations (Any answer based on landmass distribution)? The high pressure regions over oceans are also referred to as *subtropical highs*.

January map: The *Icelandic low* and the *Aleutian low* are positioned over the North Atlantic Ocean and the North Pacific respectively, whereas *North American high* and the *Siberian high* over Eurasia are distinctly seen over dominant landmasses. The high pressure areas develop due to intense cooling of the continents and low pressure regions locate their position over warm ocean waters. The airmasses (defined as the volume of air of uniform temperature and density) over Icelandic and Aleutian lows are relatively warm and moist as compared to the cool dry continental airmasses. One may infer weather development over a particular region even from the mean charts by assessing how the airmasses of different origin move on the chart. The highs and lows over the globe are quasi-permanent (and change only with the season); consequently, the associated large scale circulation is quasi-steady. In the northern hemisphere, wind blows anticlockwise around a low pressure system and clockwise around a high pressure system. , under the influence of Aleutian low (1002 hPa), warm marine air masses move from Pacific Ocean to west coast of North America. However, under the influence of stronger Icelandic low (996 hPa), cold marine air moves over east coast of America, so the airmass is cold and moist. Besides under variable situations, North America could also receive cold moist airmass from Pacific Ocean; severely cold dry Polar or Arctic airmass from north; and warm moist airmass from the south. But, airmasses do not mix but retain their characteristic boundaries during their movement, so it is possible to predict weather at a place by assessing what type of airmasses move over an area and how long that area remains under their influence. Thus, rains, fronts and snowfalls dominate the weather in winters over North America while the Indian subcontinent is dominated by the winter monsoon and western disturbances.

The characteristic features of northern winter seen in Figure are as follows:

(i)   Two low pressure systems: Aleutian Low and Icelandic Low;

(ii) Four semi-permanent highs: Siberian high and the North American high (both shallow) and two high pressure regions over oceans;

(iii) ITCZ shifted to the southern hemisphere;

(iv) Semi-permanent high pressure regions over oceans in the Southern Hemisphere;

(v)   Low pressure regions are zones of cyclonic activity where numerous storms tend to converge after travelling northeast;

(vi) Winter monsoon over India.

January mean sea-level pressure distribution

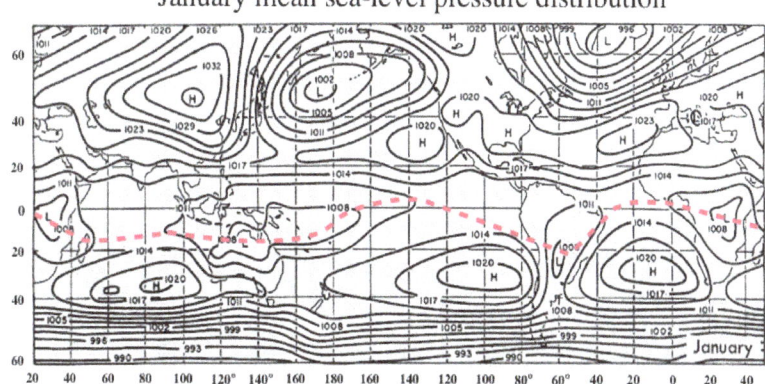

January mean sea-level pressure prepared by Y. Mintz and G. Dean (1952).
Note the position of Lows and highs in the two hemispheres.

July map: The mean sea level pressures and the associated surface circulation of July are shown in Figure, which shows that the *Icelandic low* has moved up north with centre over southern Greenland and the *Siberian high* has vanished. It is replaced by a strong thermal low over Iran, Pakistan and northwest India. The semi-permanent highs in the southern hemisphere are almost positioned over the same locations. Besides, one may notice a high pressure region on the Australian land mass which was down south over the ocean in the January map. Interestingly, the changes from winter to summer are much spectacular in the northern hemisphere as compared to the southern hemisphere. Since northern hemisphere has dominant landmasses, as summer approaches these land masses are heated up, the shallow highs vanish and thermal lows form in these regions. The thermal low over the desert region of southwest (California) may also be noted in the July map.

One of the most spectacular things that happens as a result of strengthening of the heat low over Iran, northwest India and Pakistan region (30°N, 90°E) is thearrival of warm moisture-laden winds of marine origin over the Indian subcontinent, which produce copious rains from June to September every year. There is sudden jump in the quantum of all-India daily rainfall, and more so, such a quantum jump in the mean daily rainfall is nowhere noticed on the globe. Over the oceans one may note high pressure centres viz., the Pacific high and the high pressure region over the Azores (also known as Azores or Bermuda high). The characteristic features of northern summer in Figure are as follows:

- Formation of thermal (heat) lows over Iran-Pakistan-northwest India and over the south-west United States of America;

- The dominant monsoon circulation drawing moist marine winds over land producing co-pious rains over India;

- ITCZ: its average location is in the northern hemisphere (sometimes double ITCZ – one over land and other over the ocean).

The heat low and the spectacular advance of wet monsoon are closely inter-related as the strong pressure gradient (created with the presence high pressure over the Indian Ocean) and Coriolis force together produce strong moisture-laden cross equatorial flow from southern to northern hemisphere. This inter-hemispheric exchange gives rise to moisture surge over the Indian Penin-sula and a quantum jump in the seasonal rainfall in the region.

July mean sea-level pressure distribution

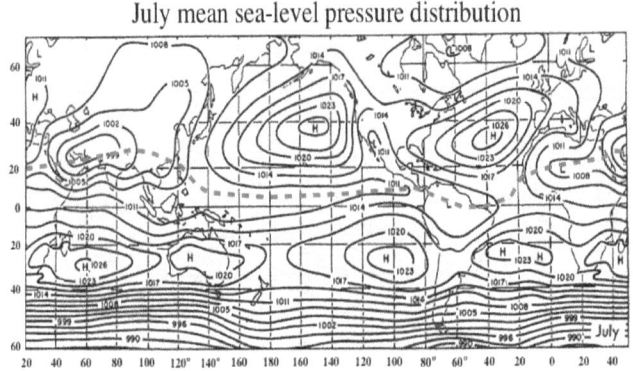

July mean sea-level pressure prepared by Y. Mintz and G. Dean (1952). Note that the position of Lows and highs in the Southern Hemisphere is practically unchanged

Indian Ocean Circulation

In oceanography, currents are always expressed with reference to the oceanic floor (rotating with the earth) where relative velocity is zero and effect of rotation is accounted for in the balances of forces. However, if origin be fixed at the sea surface, it should not cause any confusion while in-terpreting the currents increasing or decreasing with depth in agreement with the thermal wind relation in the ocean where geostrophic balance holds. The surface currents are mainly driven by winds, but countercurrents especially over the equator are also produced in the ocean. Wind driven currents and the countercurrents have been discovered in the global ocean and have been successfully explained and depicted from a long series of data by Sverdrup (1947), Stommel (1948) and Munk (1950). The global ocean circulation will be discussed later. Here, we take a glimpse of the surface currents in the tropical Indian Ocean (Shenoi et al. J. Marine Res., vol.57, 1999); these currents change in response to the annually reversing monsoon winds.In the Indian Ocean both warm and cold currents exist which form an equatorial gyre in summer.

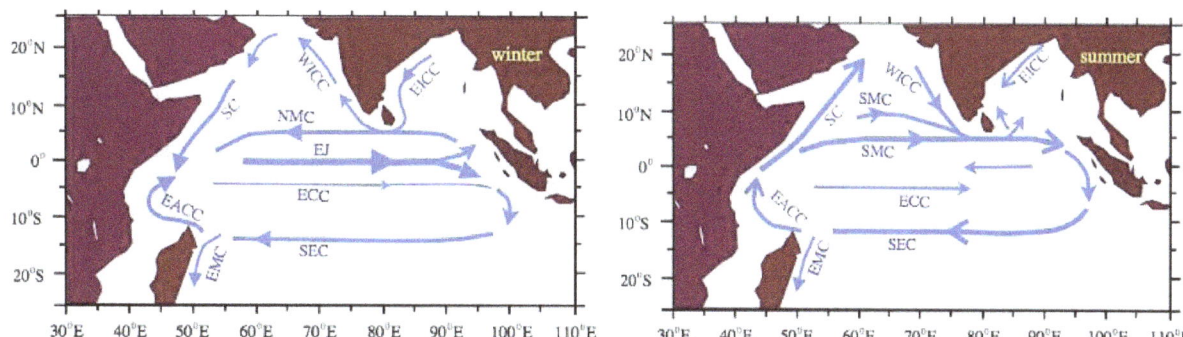

Schematic of major surface currents in the Indian Ocean during winter and summer: (a) the northeast monsoon and (b) the southwest monsoon. The major currents depicted are: South Equatorial Current (SEC), Northeast Monsoon Current (NMC), Equatorial Counter Current (ECC), Equatorial Jet (EJ), East African Coastal Current (EACC), Somali Current (SC), Southwest Monsoon Current (SMC), West India Coastal Current (WICC), East India Coastal Current (EICC) and East Madagascar Current (EMC). The EJ, though depicted in the schematic for winter, does not appear either during summer or winter monsoon season; it appears during the transition period in April–May and November–December. The thickness of the curve represents the relative magnitude of the current.

Another important feature of the surface currents of the Indian Ocean is the appearance of Equatorial Jet during the transition periods before summer (April-May) and winter (November-December). The currents in the North Indian Ocean show greater response to the reversing monsoon winds. The Somali Current is a western boundary current forced by the monsoon time Somali jet at 850 *hPa* over the Arabian Sea. The description of the currents in the Indian Ocean given here is based on an exhaustive study of Shenoi et al. (1999) that documents these currents using data from satellite tracked drifting buoys and also presents a concise summary of these currents as given in Table below.

Table: Major surface currents in the tropical Indian Ocean		
(Ref: SSC Shenoi, PK Saji and AM Almeida, Journal of Marine Research, 57, 885–907, 1999)		
Name of Current	**Description**	**Remarks**
South Equatorial Current (SEC)	A westward flow between 8°S and 16°S extending from 95°E to 50°E. Exists throughout the year.	
East Madagascar Current (EMC)	A southward flow, fed by the SEC, along the east coast of Madagascar. Exists throughout the year.	
East Africa Coastal Current (EACC)	A northward flow, fed by the SEC, along the east coast of Africa. Exists throughout the year.	
Equatorial Counter Current (ECC)	An eastward flow to the south (between 3°S to 5°S) of the equator. Absent during August.	Also known as South Equatorial Counter Current (SECC)
Equatorial Jet (EJ)	An intense eastward current that appears at the equator twice a year (during April–May and November-December).	Also known as Wyrtki Jet
Northeast Monsoon Current (NMC)	A westward flow to the north of the equator, with its axis around 5°N; exists only during December–April.	Also known as North Equatorial Current (NEC)

Southwest Monsoon Current (SMC)	An eastward flow to the north of theequator, with its axis around 5°N. Develops in June, associated with the south west monsoon winds over the north Indian Ocean, and persists till October.	Also known as Indian Monsoon Current (IMC)
Somali Current (SC)	Flows southward along the coast of Somalia during December–February and flows northward during March–September.	
West India Coastal Current (WICC)	The current along the east coast of India, flows during February–April poleward and equatorward during November–December. During the southwest monsoon a weak poleward flow develops in the south (south of 15°N) and an equatorward flow develops in the north.	
East India Coastal Current (EICC)	The current along the east coast of India, flows during February–April poleward and equatorward during November–December. During the southwest monsoon a weak poleward flow develops in the south (south of 15°N) and an equatorward flow develops in the north.	

The surface currents show mesoscale variability and their typical speed is below 50 cm s⁻¹ but it could be as high as 100 cm s⁻¹ on certain occasions on the west. The mesoscale variability arises from mesoscale processes such as ocean eddies, isolated turbulent vortices and meandering currents which have spatial scales 50-500 km and temporal scales 10-100 days. From Figure. one may observe such variability in the sea surface height (SSH). Since water masses characterized by the temperature-salinity diagram are related to winds or ocean currents, or are formed locally, the movement of water masses is linked to monsoon circulation. Moreover, stronger currents make a major impact on the transport of heat and salinity in this basin. The major surface currents in the Indian Ocean are shown in Figure that change seasonally more in the north as compared to south in the Indian Ocean. One of the interesting features of these currents is that most of them, especially of north and equatorial Indian Ocean, reverse in direction with season. However, a high speed surface current that flows eastward, the so-called Equatorial Jet (EJ), appears during both the transition periods (April–May and October-November) between the monsoons. The EJ has been shown with the wintertime current systems in Figure and this time varying current can produce profound changes in the in the mass structure of the Indian Ocean.

One may also observe in Figure the two counter rotating gyres during wintertime: (i) a southern clockwise rotating gyre with the South Equatorial Current (SEC) and the Equatorial Countercurrent (ECC); and (ii) a northern counterclockwise rotating gyre bounded by the EJ on the south and NMC on the north. But the two-gyre system breaks down during the boreal (northern) summer and only a single large clockwise rotating gyre may be observed during summertime, which is bounded by SEC on the south and SMC on the north. During summer, the SEC moves nearer to the equator. The characterization of the surface circulation of the Indian Ocean as a two-gyre system was earlier discussed by Molinari et al. (1990). An interesting feature of the wintertime surface circulation of the Indian Ocean is that it also resembles the general circulations of Pacific and Atlantic oceans, but not the summertime circulation.

In a recent review paper entitled *Indian ocean circulation and climate variability by* Schott et al. an update on the identified current systems during winter and summer in the Indian Ocean has presented alongwith the Indian Ocean Dipole in the sea surface temperatures. The key finding of recent studies is that the sea surface temperatures and the heat content of the Indian Ocean have been increasing over the past decades.

If the oceans were at rest covering the entire earth, then the sea surface would be a geopotential surface (i.e. moving along this would do no work against gravity); it is therefore referred to as the *geoid* of the earth. Any deviations from the geoid would introduce slopes in the sea surface, which can be measured if the geoid is known. The changes in the sea surface height (or steric height) are caused by currents (i.e. ocean dynamics) and therefore, it is also referred to as the *dynamic topography* of the ocean. The mean sea surface topography or long time averaged sea surface includes the geoid height. To measure the dynamic topography of the order of a few centimetres, accurate radar altimeter satellites are required. Here we have shown in Figure the dynamic height of the Indian Ocean from TOPEX/Poseidon Satellite for the period 1-11 May 1994. The analysis of Shenoi et al. (1999) also shows: (i) currents calculated from dynamic topography (0/1000 db) may not always represent correctly the surface circulation; and (ii) turbulent eddies dominate the circulation which has been inferred from the eddy kinetic energy with its source at the western boundary (along the coast of Africa) and sinks in Arabian Sea and Bay of Bengal in the north and regions south of 20°S.

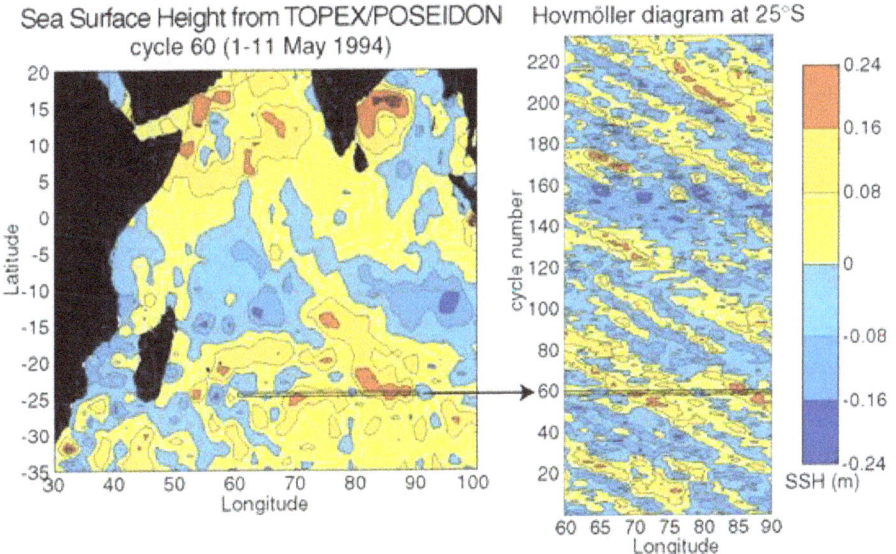

Sea surface height (SSH) of the Indian Ocean and Hovmöller diagram at 25°S

The Hovmöller diagram of the SSH shows the Rossby waves propagating westward. This requires accurate steric height estimation from satellites because various kinds of studies of the oceans can be performed with this data. Further, SSH and the pressure are related, therefore the height field is also related to the velocity field. Like in the atmosphere, balance of pressure gradient force, Coriolis force and the frictional force determines the kind of circulation that would dominate the ocean depths. Indeed, the time evolution of ocean circulation depends of the external factors, viz., wind stress, heating and cooling, and evaporation and precipitation. Hence, the ocean temperature, salinity and other properties depend on these three external factors and also how the moving seawater parcels would balance under the combined action of these forces.

The general circulation of the global ocean is anticyclonic, which implies that circulation is clockwise in the northern hemisphere and counter clockwise in the southern hemisphere. The observed ocean currents as shown in the Figure are a combination of geostrophic current and those forced by wind stress. The names of these currents are also mentioned on the figure. The ocean circulation differs in the tropics, subtropics and the subpolar latitudes. The most spectacular feature of the ocean circulation from Figure is that it is dominated by *gyres* (closed circulations), which have anticyclonic orientation in the Pacific and Atlantic Oceans. A gyre in the Indian Ocean is present during summer but it splits into two during winter. The current speed in the gyres is about 10 cm s⁻¹ but on the western edge of the gyre the currents could attain the speed as high as 100 cm s⁻¹ or more and are known as the western boundary currents. The northward flowing currents, viz., the Gulf Stream in the north Atlantic and Kuroshio in the North Pacific, are the western boundary currents. The typical current speeds on the eastern edge of ocean gyres are much smaller than that on its western edge; this gives rise to a strong east-west asymmetry in the current system.

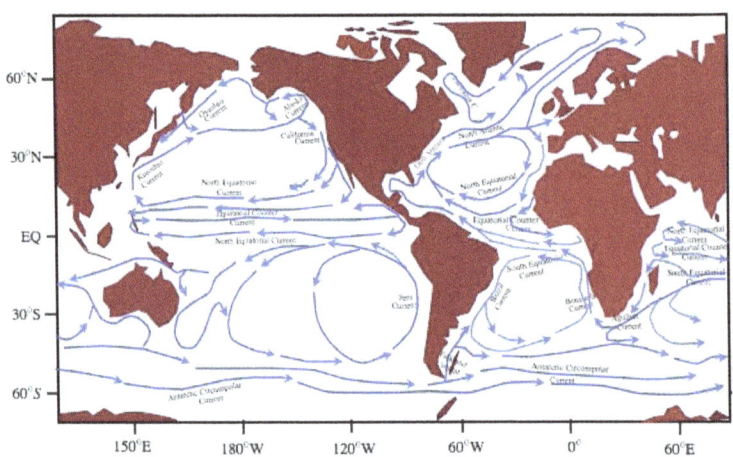

The general pattern of ocean currents: combination of geostrophic and wind driven currents. Note that circulations are anticyclonic.

The zonal currents in the northern hemisphere are blocked by the continental boundaries. Also the surface currents reconstituted by the trajectory of drifters reveal that eddies dominate the global oceans. Both in the Atlantic and Pacific, the currents in the tropics are westward and a counter current flows on the equator. The wintertime currents in the Indian Ocean is very similar to this type of current system, but the southwest monsoon produces the Somali current which, in turn, gives rise upwelling of cold water off the east coast of Somalia. The Somali current is also a northward flowing western boundary current. In the subpolar regions of Atlantic and Pacific, the gyres are cyclonic. The most notable effect of the western boundary currents is the strong temperature gradients which they produce in the interior of the ocean, though they weaken with the depth. In the tropics, temperature gradient is stronger in the zonal direction as compared to the meridional component. Another important feature of the ocean circulation is that the Antarctic Circumpolar Current dominates the Southern Ocean and it is not blocked anywhere by the coastal boundaries.

Weather Station

Weather station at Mildura Airport, Victoria, Australia.

A weather station is a facility, either on land or sea, with instruments and equipment for measuring atmospheric conditions to provide information for weather forecasts and to study the weather and climate. The measurements taken include temperature, atmospheric pressure, humidity, wind speed, wind direction, and precipitation amounts. Wind measurements are taken with as few other obstructions as possible, while temperature and humidity measurements are kept free from direct solar radiation, or insolation. Manual observations are taken at least once daily, while automated measurements are taken at least once an hour. Weather conditions out at sea are taken by ships and buoys, which measure slightly different meteorological quantities such as sea surface temperature (SST), wave height, and wave period. Drifting weather buoys outnumber their moored versions by a significant amount.

Instruments

The NOAA weather station at Wake Island harbor measures and transmits data on wind speed, atmospheric pressure, air temperature and tides.

Typical weather stations have the following instruments:

- Thermometer for measuring air and sea surface temperature

- Barometer for measuring atmospheric pressure

- Hygrometer for measuring humidity

- Anemometer for measuring wind speed

- Pyranometer for measuring solar radiation

- Rain gauge for measuring liquid precipitation over a set period of time.

In addition, at certain Automated airport weather stations, additional instruments may be employed, including:

- Present Weather/Precipitation Identification Sensor for identifying falling precipitation

- Disdrometer for measuring drop size distribution

- Transmissometer for measuring visibility

- Ceilometer for measuring cloud ceiling

More sophisticated stations may also measure the ultraviolet index, leaf wetness, soil moisture, soil temperature, water temperature in ponds, lakes, creeks, or rivers, and occasionally other data.

Exposure

Except for those instruments requiring direct exposure to the elements (anemometer, rain gauge), the instruments should be sheltered in a vented box, usually a Stevenson screen, to keep direct sunlight off the thermometer and wind off the hygrometer. The instrumentation may be specialized to allow for periodic recording otherwise significant manual labour is required for record keeping. Automatic transmission of data, in a format such as METAR, is also desirable as many weather station's data is required for weather forecasting.

Personal Weather Station

A personal weather station is a set of weather measuring instruments operated by a private individual, club, association, or even business (where obtaining and distributing weather data is not a part of the entity's business operation). The quality, number of instruments, and placement of personal weather stations can vary widely, making the determination of which stations collect accurate, meaningful, and comparable data difficult.

Personal weather stations also typically involve a digital console that provides readouts of the data being collected. These consoles may interface to a personal computer where data can be displayed, stored, and uploaded to websites or data ingestion/distribution systems. Open source weather stations are available that are designed to be fully customizable by users.

Roof-mounted weather station instruments

Personal weather stations may be operated solely for the enjoyment and education of the owner, however some owners share their results with others. They do this by either by manually compiling data and distributing it, distributing data over the internet, or sharing data via amateur radio. The Citizen Weather Observer Program (CWOP) is a service which facilitates the sharing of information from personal weather stations. This data is submitted through use of software, a personal computer, and internet connection (or amateur radio) and are utilized by groups such as the National Weather Service (NWS) when generating forecast models. Each weather station submitting data to CWOP will also have an individual Web page that depicts the data submitted by that station. The Weather Underground Internet site is another popular destination for the submittal and sharing of data with others around the world. As with CWOP, each station submitting data to Weather Underground has a unique Web page displaying their submitted data. The UK Met Office's Weather Observations Website (WOW) also allows such data to be shared and displayed.

The weather ship M/S *Polarfront* at sea.

Dedicated Ships

A weather ship was a ship stationed in the ocean as a platform for surface and upper air meteorological measurements for use in weather forecasting. It was also meant to aid in search and rescue operations and to support transatlantic flights. The establishment of weather ships proved to be so useful during World War II that the International Civil Aviation Organization (ICAO) established a global network of 13 weather ships in 1948. Of the 12 left in operation in 1996, nine were located in the northern Atlantic ocean while three were located in the northern Pacific ocean. The agreement of the weather ships ended in 1990. Weather ship observations proved to be helpful in wind and wave studies, as they did not avoid weather systems like merchant ships tended to and were considered a valuable resource. The last weather ship was MS *Polarfront*, known as weather station M ("jilindras") at 66°N, 02°E, run by the Norwegian Meteorological Institute. MS *Polarfront* was removed from service January 1, 2010. Since the 1960s this role has been largely superseded by satellites, long range aircraft and weather buoys. Weather observations from ships continue from thousands of voluntary merchant vessels in routine commercial operation; the Old Weather crowdsourcing project transcribes naval logs from before the era of dedicated ships.

Weather buoy operated by the NOAA National Data Buoy Center

Dedicated Buoys

Weather buoys are instruments which collect weather and oceanography data within the world's oceans and lakes. Moored buoys have been in use since 1951, while drifting buoys have been used since the late 1970s. Moored buoys are connected with the seabed using either chains, nylon, or buoyant polypropylene. With the decline of the weather ship, they have taken a more primary role in measuring conditions over the open seas since the 1970s. During the 1980s and 1990s, a network of buoys in the central and eastern tropical Pacific ocean helped study the El Niño-Southern Oscillation. Moored weather buoys range from 1.5 metres (4.9 ft) to 12 metres (39 ft) in diameter, while drifting buoys are smaller, with diameters of 30 centimetres (12 in) to 40 centimetres (16 in). Drifting buoys are the dominant form of weather buoy in sheer number, with 1250 located worldwide. Wind data from buoys has smaller error than that from ships. There are differences in the values of sea surface temperature measurements between the two platforms as well, relating to the depth of the measurement and whether or not the water is heated by the ship which measures the quantity.

Synoptic Weather Station

Synoptic weather stations are instruments which collect meteorological information at synoptic time 00h00, 06h00, 12h00, 18h00 (UTC) and at intermediate synoptic hours 03h00, 09h00, 15h00, 21h00 (UTC).

The common instruments of measure are anemometer, wind vane, pressure sensor, thermometer, hygrometer, and rain gauge.

The weather measures are formatted in special format and transmit to WMO to help the weather forecast model.

Synoptic Automatic Weather Station

Networks

A variety of land-based weather station networks have been set up globally. Some of these are basic to analyzing weather fronts and pressure systems, such as the synoptic observation network, while others are more regional in nature, known as mesonets.

Global

- Citizen Weather Observer Program (CWOP)
- Weather Underground Personal Weather Stations

United States

- Arizona Meteorological Network (AZMET)

- Central Pennsylvania Volunteer Weather Station Network

- Florida Automated Weather Network (FAWN)

- Georgia Environmental Monitoring Network (GAEMN)

- Indiana Purdue Automated Agricultural Weather Station Network (PAAWS)

- Iowa Environmental Mesonet (IEM)

- MesoWest

- Michigan Automated Weather Network (MAWN)

- Missouri Weather Stations

- National Weather Service Cooperative Observer (COOP) program

- Oklahoma Mesonet

- The Pacific Northwest Cooperative Agricultural Weather Network

Southern Hemisphere

- Antarctic Automatic Weather Stations Project

- Australia/Bureau of Meteorology AWS network.

- Australia/Department of Agriculture and Food Western Australia

- Australia/Lower Murray Water Automatic Weather Station Network

Weather Forecasting

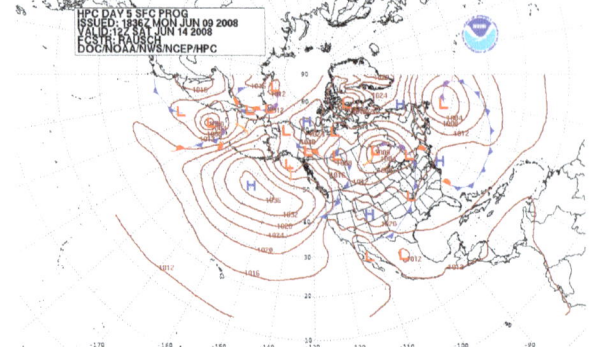

Forecast of surface pressures five days into the future for the north Pacific,
North America, and north Atlantic Ocean

Weather forecasting is the application of science and technology to predict the state of the atmosphere for a given location. Human beings have attempted to predict the weather informally for millennia, and formally since the nineteenth century. Weather forecasts are made by collecting quantitative data about the current state of the atmosphere at a given place and using scientific understanding of atmospheric processes to project how the atmosphere will change.

Once an all-human endeavor based mainly upon changes in barometric pressure, current weather conditions, and sky condition, weather forecasting now relies on computer-based models that take many atmospheric factors into account. Human input is still required to pick the best possible forecast model to base the forecast upon, which involves pattern recognition skills, teleconnections, knowledge of model performance, and knowledge of model biases. The inaccuracy of forecasting is due to the chaotic nature of the atmosphere, the massive computational power required to solve the equations that describe the atmosphere, the error involved in measuring the initial conditions, and an incomplete understanding of atmospheric processes. Hence, forecasts become less accurate as the difference between current time and the time for which the forecast is being made (the *range* of the forecast) increases. The use of ensembles and model consensus help narrow the error and pick the most likely outcome.

There are a variety of end uses to weather forecasts. Weather warnings are important forecasts because they are used to protect life and property. Forecasts based on temperature and precipitation are important to agriculture, and therefore to traders within commodity markets. Temperature forecasts are used by utility companies to estimate demand over coming days. On an everyday basis, people use weather forecasts to determine what to wear on a given day. Since outdoor activities are severely curtailed by heavy rain, snow and wind chill, forecasts can be used to plan activities around these events, and to plan ahead and survive them. In 2014, the US spent $5.1 billion on weather forecasting.

History

Ancient Forecasting

For millennia people have tried to forecast the weather. In 650 BC, the Babylonians predicted the weather from cloud patterns as well as astrology. In about 350 BC, Aristotle described weather patterns in *Meteorologica*. Later, Theophrastus compiled a book on weather forecasting, called the *Book of Signs*. Chinese weather prediction lore extends at least as far back as 300 BC, which was also around the same time ancient Indian astronomers developed weather-prediction methods. In New Testament times, Christ himself referred to deciphering and understanding local weather patterns, by saying, "When evening comes, you say, 'It will be fair weather, for the sky is red', and in the morning, 'Today it will be stormy, for the sky is red and overcast.' You know how to interpret the appearance of the sky, but you cannot interpret the signs of the times."

In 904 AD, Ibn Wahshiyya's *Nabatean Agriculture* discussed the weather forecasting of atmospheric changes and signs from the planetary astral alterations; signs of rain based on observation of the lunar phases; and weather forecasts based on the movement of winds.

Ancient weather forecasting methods usually relied on observed patterns of events, also termed pattern recognition. For example, it might be observed that if the sunset was particularly red, the following day often brought fair weather. This experience accumulated over the generations to produce weather lore. However, not all of these predictions prove reliable, and many of them have since been found not to stand up to rigorous statistical testing.

Modern Methods

It was not until the invention of the electric telegraph in 1835 that the modern age of weather forecasting began. Before that, the fastest that distant weather reports could travel was around 100 miles per day (160 km/d), but was more typically 40–75 miles per day (60–120 km/day) (whether by land or by sea). By the late 1840s, the telegraph allowed reports of weather conditions from a wide area to be received almost instantaneously, allowing forecasts to be made from knowledge of weather conditions further upwind.

The *Royal Charter* sank in an 1859 storm, stimulating the establishment of modern weather forecasting.

The two men credited with the birth of forecasting as a science were officer of the Royal Navy Francis Beaufort and his protégé Robert FitzRoy. Both were influential men in British naval and governmental circles, and though ridiculed in the press at the time, their work gained scientific credence, was accepted by the Royal Navy, and formed the basis for all of today's weather forecasting knowledge.

Beaufort developed the Wind Force Scale and Weather Notation coding, which he was to use in his journals for the remainder of his life. He also promoted the development of reliable tide tables around British shores, and with his friend William Whewell, expanded weather record-keeping at 200 British Coast guard stations.

Robert FitzRoy was appointed in 1854 as chief of a new department within the Board of Trade to deal with the collection of weather data at sea as a service to mariners. This was the forerunner of the modern Meteorological Office. All ship captains were tasked with collating data on the weather and computing it, with the use of tested instruments that were loaned for this purpose.

A storm in 1859 that caused the loss of the *Royal Charter* inspired FitzRoy to develop charts to allow predictions to be made, which he called *"forecasting the weather"*, thus coining the term "weather forecast". Fifteen land stations were established to use the new telegraph to transmit to him daily reports of weather at set times leading to the first gale warning service. His warning service for shipping was initiated in February 1861, with the use of telegraph communications. The first daily weather forecasts were published in *The Times* in 1861. In the following year a system was introduced of hoisting storm warning cones at the principal ports when a gale was expected. The *"Weather Book"* which FitzRoy published in 1863 was far in advance of the scientific opinion of the time.

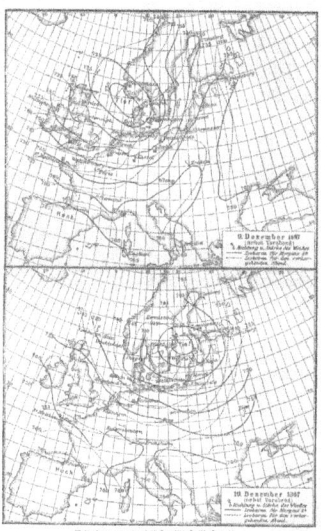

Weather map of Europe, 10 December 1887.

As the electric telegraph network expanded, allowing for the more rapid dissemination of warnings, a national observational network was developed which could then be used to provide synoptic analyses. Instruments to continuously record variations in meteorological parameters using photography were supplied to the observing stations from Kew Observatory – these cameras had been invented by Francis Ronalds in 1845 and his barograph had earlier been used by FitzRoy.

To convey accurate information, it soon became necessary to have a standard vocabulary describing clouds; this was achieved by means of a series of classifications first achieved by Luke Howard in 1802, and standardized in the *International Cloud Atlas* of 1896.

Numerical Prediction

It was not until the 20th century that advances in the understanding of atmospheric physics led to the foundation of modern numerical weather prediction. In 1922, English scientist Lewis Fry Richardson published "Weather Prediction By Numerical Process", after finding notes and derivations he worked on as an ambulance driver in World War I. He described therein how small terms in the prognostic fluid dynamics equations governing atmospheric flow could be neglected, and a finite differencing scheme in time and space could be devised, to allow numerical prediction solutions to be found.

Richardson envisioned a large auditorium of thousands of people performing the calculations and passing them to others. However, the sheer number of calculations required was too large to be completed without the use of computers, and the size of the grid and time steps led to unrealistic results in deepening systems. It was later found, through numerical analysis, that this was due to numerical instability. The first computerised weather forecast was performed by a team led by the mathematician John von Neumann; von Neumann publishing the paper *Numerical Integration of the Barotropic Vorticity Equation* in 1950. Practical use of numerical weather prediction began in 1955, spurred by the development of programmable electronic computers.

Broadcasts

George Cowling (above) presented the first in-vision forecast on 11 January 1954 for the BBC.

The first ever daily weather forecasts were published in *The Times* on 1 August 1861, and the first weather maps were produced later in the same year. In 1911, the Met Office began issuing the first marine weather forecasts via radio transmission. These included gale and storm warnings for areas around Great Britain. In the United States, the first public radio forecasts were made in 1925 by Edward B. "E.B." Rideout, on WEEI, the Edison Electric Illuminating station in Boston. Rideout came from the U.S. Weather Bureau, as did WBZ weather forecaster G. Harold Noyes in 1931.

The world's first televised weather forecasts, including the use of weather maps, were experimentally broadcast by the BBC in 1936. This was brought into practice in 1949 after World War II. George Cowling gave the first weather forecast while being televised in front of the map in 1954. In America, experimental television forecasts were made by James C Fidler in Cincinnati in either 1940 or 1947 on the DuMont Television Network. In the late 1970s and early 80s, John Coleman, the first weatherman on ABC-TV's Good Morning America, pioneered the use of on-screen weather satellite information and computer graphics for television forecasts. Coleman was a co-founder of The Weather Channel (TWC) in 1982. TWC is now a 24-hour cable network. Some weather channels have started broadcasting on live broadcasting programs such as YouTube and Periscope to reach more viewers. One Meteorologist, Jeff Matthews, transformed the medium from forecasting live during television broadcasts to forecasting live on Periscope under the name of "Weathertainment", becoming the first meteorologist to deliver daily forecasts on a life-streaming application... and thus, making weather forecasting, interactive, for the public.

How Models Create Forecasts

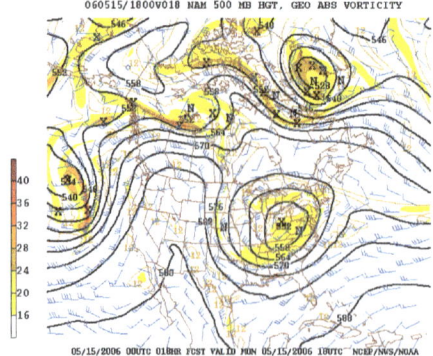

An example of 500 mbar geopotential height and absolute vorticity prediction from a numerical weather prediction model

The basic idea of numerical weather prediction is to sample the state of the fluid at a given time and use the equations of fluid dynamics and thermodynamics to estimate the state of the fluid at some time in the future. The main inputs from country-based weather services are surface observations from automated weather stations at ground level over land and from weather buoys at sea. The World Meteorological Organization acts to standardize the instrumentation, observing practices and timing of these observations worldwide. Stations either report hourly in METAR reports, or every six hours in SYNOP reports. Sites launch radiosondes, which rise through the depth of the troposphere and well into the stratosphere. Data from weather satellites are used in areas where traditional data sources are not available. Compared with similar data from radiosondes, the satellite data has the advantage of global coverage, however at a lower accuracy and resolution. Meteorological radar provide information on precipitation location and intensity, which can be used to estimate precipitation accumulations over time. Additionally, if a pulse Doppler weather radar is used then wind speed and direction can be determined.

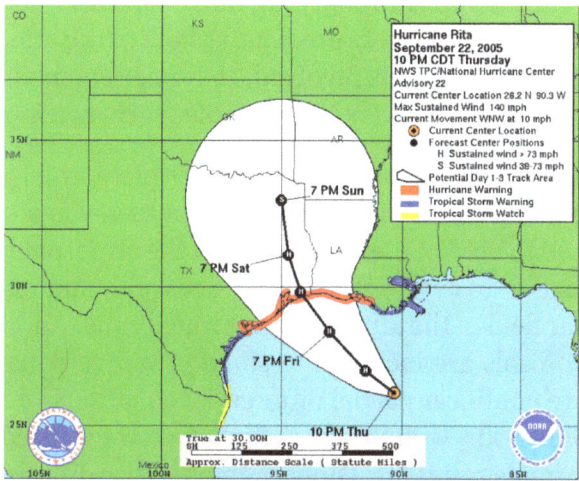

Modern weather predictions aid in timely evacuations and potentially save lives and prevent property damage

Commerce provides pilot reports along aircraft routes, and ship reports along shipping routes. Research flights using reconnaissance aircraft fly in and around weather systems of interest such as tropical cyclones. Reconnaissance aircraft are also flown over the open oceans during the cold season into systems which cause significant uncertainty in forecast guidance, or are expected to be of high impact 3–7 days into the future over the downstream continent.

Models are *initialized* using this observed data. The irregularly spaced observations are processed by data assimilation and objective analysis methods, which perform quality control and obtain values at locations usable by the model's mathematical algorithms (usually an evenly spaced grid). The data are then used in the model as the starting point for a forecast. Commonly, the set of equations used to predict the known as the physics and dynamics of the atmosphere are called primitive equations. These equations are initialized from the analysis data and rates of change are determined. The rates of change predict the state of the atmosphere a short time into the future. The equations are then applied to this new atmospheric state to find new rates of change, and these new rates of change predict the atmosphere at a yet further time into the future. This *time stepping* procedure is continually repeated until the solution reaches the desired forecast time. The length of the time step is related to the distance between the points on the computational grid.

The length of the time step chosen within the model is related to the distance between the points on the computational grid, and is chosen to maintain numerical stability. Time steps for global models are on the order of tens of minutes, while time steps for regional models are between one and four minutes. The global models are run at varying times into the future. The Met Office's Unified Model is run six days into the future, the European Centre for Medium-Range Weather Forecasts model is run out to 10 days into the future, while the Global Forecast System model run by the Environmental Modeling Center is run 16 days into the future. The visual output produced by a model solution is known as a prognostic chart, or *prog*. The raw output is often modified before being presented as the forecast. This can be in the form of statistical techniques to remove known biases in the model, or of adjustment to take into account consensus among other numerical weather forecasts. MOS or model output statistics is a technique used to interpret numerical model output and produce site-specific guidance. This guidance is presented in coded numerical form, and can be obtained for nearly all National Weather Service reporting stations in the United States. As proposed by Edward Lorenz in 1963, long range forecasts, those made at a range of two weeks or more, are impossible to definitively predict the state of the atmosphere, owing to the chaotic nature of the fluid dynamics equations involved. In numerical models, extremely small errors in initial values double roughly every five days for variables such as temperature and wind velocity.

Essentially, a model is a computer program that produces meteorological information for future times at given locations and altitudes. Within any modern model is a set of equations, known as the primitive equations, used to predict the future state of the atmosphere. These equations—along with the ideal gas law—are used to evolve the density, pressure, and potential temperature scalar fields and the velocity vector field of the atmosphere through time. Additional transport equations for pollutants and other aerosols are included in some primitive-equation mesoscale models as well. The equations used are nonlinear partial differential equations which are impossible to solve exactly through analytical methods, with the exception of a few idealized cases. Therefore, numerical methods obtain approximate solutions. Different models use different solution methods: some global models use spectral methods for the horizontal dimensions and finite difference methods for the vertical dimension, while regional models and other global models usually use finite-difference methods in all three dimensions.

Techniques

Persistence

The simplest method of forecasting the weather, persistence, relies upon today's conditions to forecast the conditions tomorrow. This can be a valid way of forecasting the weather when it is in a steady state, such as during the summer season in the tropics. This method of forecasting strongly depends upon the presence of a stagnant weather pattern. Therefore, when in a fluctuating weather pattern, this method of forecasting becomes inaccurate. It can be useful in both short range forecasts and long range forecasts.

Use of a Barometer

Measurements of barometric pressure and the pressure tendency (the change of pressure over time) have been used in forecasting since the late 19th century. The larger the change in pressure, especially if more than 3.5 hPa (2.6 mmHg), the larger the change in weather can be expected. If the pressure drop is rapid, a low pressure system is approaching, and there is a greater chance of rain. Rapid pressure rises are associated with improving weather conditions, such as clearing skies.

Looking at the Sky

Along with pressure tendency, the condition of the sky is one of the more important parameters used to forecast weather in mountainous areas. Thickening of cloud cover or the invasion of a higher cloud deck is indicative of rain in the near future. At night, high thin cirrostratus clouds can lead to halos around the moon, which indicates an approach of a warm front and its associated rain. Morning fog portends fair conditions, as rainy conditions are preceded by wind or clouds which prevent fog formation. The approach of a line of thunderstorms could indicate the approach of a cold front. Cloud-free skies are indicative of fair weather for the near future. A bar can indicate a coming tropical cyclone. The use of sky cover in weather prediction has led to various weather lore over the centuries.

Marestail shows moisture at high altitude, signalling the later arrival of wet weather.

Nowcasting

The forecasting of the weather within the next six hours is often referred to as nowcasting. In this time range it is possible to forecast smaller features such as individual showers and thunderstorms with reasonable accuracy, as well as other features too small to be resolved by a computer model. A human given the latest radar, satellite and observational data will be able to make a better analysis of the small scale features present and so will be able to make a more accurate forecast for the following few hours. However, there are now expert systems using those data and mesoscale numerical model to make better extrapolation, including evolution of those features in time.

Use of Forecast Models

In the past, the human forecaster was responsible for generating the entire weather forecast based upon available observations. Today, human input is generally confined to choosing a model based on various parameters, such as model biases and performance. Using a consensus of forecast models, as well as ensemble members of the various models, can help reduce forecast error. However, regardless how small the average error becomes with any individual system, large errors within any particular piece of guidance are still possible on any given model run. Humans are required to interpret the model data into weather forecasts that are understandable to the end user. Humans can use knowledge of local effects which may be too small in size to be resolved by the model to add information to the forecast. While increasing accuracy of forecast models implies that humans may no longer be needed in the forecast process at some point in the future, there is currently still a need for human intervention.

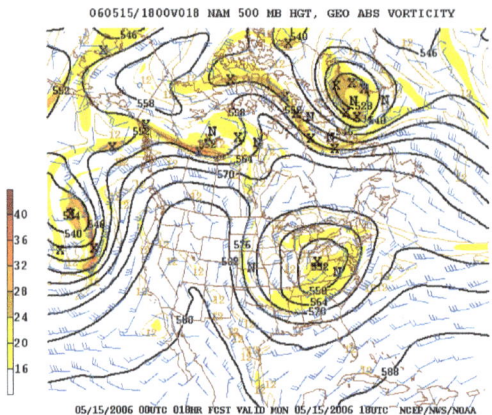

An example of 500 mbar geopotential height prediction from a numerical weather prediction model

Analog Technique

The analog technique is a complex way of making a forecast, requiring the forecaster to remember a previous weather event which is expected to be mimicked by an upcoming event. What makes it a difficult technique to use is that there is rarely a perfect analog for an event in the future. Some call this type of forecasting pattern recognition. It remains a useful method of observing rainfall over data voids such as oceans, as well as the forecasting of precipitation amounts and distribution in the future. A similar technique is used in medium range forecasting, which is known as teleconnections, when systems in other locations are used to help pin down the location of another system within the surrounding regime. An example of teleconnections are by using El Niño-Southern Oscillation (ENSO) related phenomena.

Quality Assessment

Mathematical models are often developed to map the forecast skill of a weather prediction methodology.

Communicating Forecasts to the Public

An example of a two-day weather forecast in the visual style that an American newspaper might use. Temperatures are given in Fahrenheit.

Most end users of forecasts are members of the general public. Thunderstorms can create strong winds and dangerous lightning strikes that can lead to deaths, power outages, and widespread hail damage. Heavy snow or rain can bring transportation and commerce to a stand-still, as well as cause flooding in low-lying areas. Excessive heat or cold waves can sicken or kill those with inadequate utilities, and droughts can impact water usage and destroy vegetation.

Several countries employ government agencies to provide forecasts and watches/warnings/advisories to the public in order to protect life and property and maintain commercial interests. Knowledge of what the end user needs from a weather forecast must be taken into account to present the information in a useful and understandable way. Examples include the National Oceanic and Atmospheric Administration's National Weather Service (NWS) and Environment Canada's Meteorological Service (MSC). Traditionally, newspaper, television, and radio have been the primary outlets for presenting weather forecast information to the public. Increasingly, the internet is being used due to the vast amount of specific information that can be found. In all cases, these outlets update their forecasts on a regular basis.

Severe Weather Alerts and Advisories

A major part of modern weather forecasting is the severe weather alerts and advisories which the national weather services issue in the case that severe or hazardous weather is expected. This is done to protect life and property. Some of the most commonly known of severe weather advisories are the severe thunderstorm and tornado warning, as well as the severe thunderstorm and tornado watch. Other forms of these advisories include winter weather, high wind, flood, tropical cyclone, and fog. Severe weather advisories and alerts are broadcast through the media, including radio, using emergency systems as the Emergency Alert System which break into regular programming.

Low Temperature Forecast

The low temperature forecast for the current day is calculated using the lowest temperature found between 7pm that evening through 7am the following morning. So, in short, today's forecasted low is most likely tomorrow's low temperature.

Specialist Forecasting

There are a number of sectors with their own specific needs for weather forecasts and specialist services are provided to these users.

Air Traffic

Because the aviation industry is especially sensitive to the weather, accurate weather forecasting is essential. Fog or exceptionally low ceilings can prevent many aircraft from landing and taking off. Turbulence and icing are also significant in-flight hazards. Thunderstorms are a problem for all aircraft because of severe turbulence due to their updrafts and outflow boundaries, icing due to the heavy precipitation, as well as large hail, strong winds, and lightning, all of which can cause severe damage to an aircraft in flight. Volcanic ash is also a significant problem for aviation, as aircraft can lose engine power within ash clouds. On a day-to-day basis airliners are routed to take advantage of the jet stream tailwind to improve fuel efficiency. Aircrews are briefed prior to takeoff on the conditions to expect en route and at their destination. Additionally, airports often change which runway is being used to take advantage of a headwind. This reduces the distance required for takeoff, and eliminates potential crosswinds.

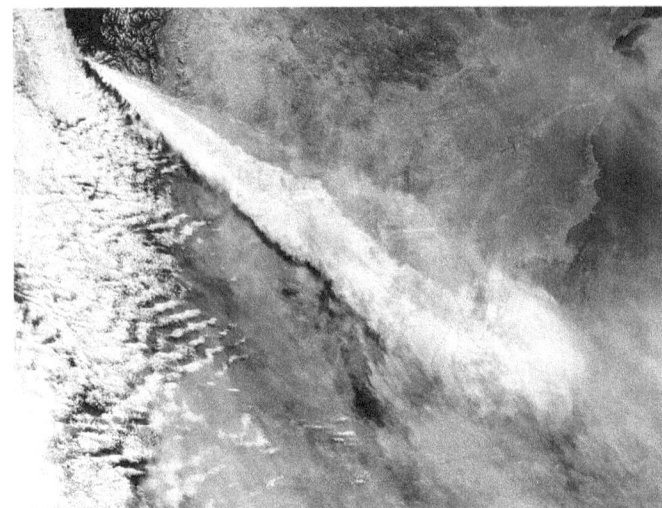

Ash cloud from the 2008 eruption of Chaitén volcano stretching across Patagonia from the Pacific to the Atlantic Ocean

Marine

Commercial and recreational use of waterways can be limited significantly by wind direction and speed, wave periodicity and heights, tides, and precipitation. These factors can each influence the safety of marine transit. Consequently, a variety of codes have been established to efficiently transmit detailed marine weather forecasts to vessel pilots via radio, for example the MAFOR (marine forecast). Typical weather forecasts can be received at sea through the use of RTTY, Navtex and Radiofax.

Agriculture

Farmers rely on weather forecasts to decide what work to do on any particular day. For example, drying hay is only feasible in dry weather. Prolonged periods of dryness can ruin cotton, wheat, and corn crops. While corn crops can be ruined by drought, their dried remains can be used as a cattle feed substitute in the form of silage. Frosts and freezes play havoc with crops both during the spring and fall. For example, peach trees in full bloom can have their potential peach crop decimated by a spring freeze. Orange groves can suffer significant damage during frosts and freezes, regardless of their timing.

Forestry

Weather forecasting of wind, precipitations and humidity is essential for preventing and controlling wildfires. Different indices, like the *Forest fire weather index* and the *Haines Index*, have been developed to predict the areas more at risk to experience fire from natural or human causes. Conditions for the development of harmful insects can be predicted by forecasting the evolution of weather, too.

Utility Companies

Electricity and gas companies rely on weather forecasts to anticipate demand which can be strongly affected by the weather. They use the quantity termed the degree day to determine how strong of a use there will be for heating (heating degree day) or cooling (cooling degree day). These quantities are based on a daily average temperature of 65 °F (18 °C). Cooler temperatures force heating degree days (one per degree Fahrenheit), while warmer temperatures force cooling degree days. In winter, severe cold weather can cause a surge in demand as people turn up their heating. Similarly, in summer a surge in demand can be linked with the increased use of air conditioning systems in hot weather. By anticipating a surge in demand, utility companies can purchase additional supplies of power or natural gas before the price increases, or in some circumstances, supplies are restricted through the use of brownouts and blackouts.

An air handling unit is used for the heating and cooling of air in a central location (click on image for legend).

Other Commercial Companies

Increasingly, private companies pay for weather forecasts tailored to their needs so that they can increase their profits or avoid large losses. For example, supermarket chains may change the stocks on their shelves in anticipation of different consumer spending habits in different weather conditions. Weather forecasts can be used to invest in the commodity market, such as futures in oranges, corn, soybeans, and oil.

Military Applications

United Kingdom Armed Forces

Royal Navy

The UK Royal Navy, working with the UK Met Office, has its own specialist branch of weather observers and forecasters, as part of the Hydrographic and Meteorological (HM) specialisation, who monitor and forecast operational conditions across the globe, to provide accurate and timely weather and oceanographic information to submarines, ships and Fleet Air Arm aircraft.

Royal Air Force

A mobile unit in the RAF, working with the UK Met Office, forecasts the weather for regions in which British, allied servicemen and women are deployed. A group based at Camp Bastion provides forecasts for the British armed forces in Afghanistan.

United States Armed Forces

US Navy

Emblem of JTWC Joint Typhoon Warning Center

US Air Force

Within the United States, Air Force Weather provides weather forecasting for the Air Force and the Army. Air Force forecasters cover air operations in both wartime and peacetime operations and provide Army support; United States Coast Guard marine science technicians provide ship forecasts for ice breakers and other various operations within their realm; and Marine forecasters provide support for ground- and air-based United States Marine Corps operations. All four military branches take their initial enlisted meteorology technical training at Keesler Air Force Base. Military and civilian forecasters actively cooperate in analyzing, creating and critiquing weather forecast products.

References

- John D. Cox (2002). Stormwatchers: The Turbulent History of Weather Prediction From Franklin's Kite to El Nino. John Wiley & Sons, Inc. pp. 53–56. ISBN 0-471-38108-X

- DataStreme Atmosphere (2008-04-28). "Air Temperature Patterns". American Meteorological Society. Archived from the original on 2008-05-11. Retrieved 2010-02-07

- Dirmeyer, Paul A.; Schlosser, C. Adam; Brubaker, Kaye L. (February 1, 2009). "Precipitation, Recycling, and Land Memory: An Integrated Analysis". JOURNAL OF HYDROMETEOROLOGY. 10: 278–288. doi:10.1175/2008JHM1016.1. Retrieved December 30, 2016

- World Meteorological Organization, ed. (1975). Étages, International Cloud Atlas (PDF). I. pp. 15–16. ISBN 92-63-10407-7. Retrieved 26 August 2014

- Allaby, Michael (2009). Atmosphere: A Scientific History of Air, Weather, and Climate. Infobase Publishing.

Retrieved 2013-12-07

- M. Fulchignoni; F. Ferri; F. Angrilli; A. Bar-Nun; M.A. Barucci; G. Bianchini; et al. (2002). "The Characterisation of Titan's Atmospheric Physical Properties by the Huygens Atmospheric Structure Instrument (Hasi)". Space Science Review. 104: 395–431. Bibcode:2002SSRv..104..395F. doi:10.1023/A:1023688607077

- Terry T. Lankford (1999). Aircraft icing: a pilot's guide. McGraw-Hill Professional. pp. 129–134. ISBN 0-07-134139-0. Retrieved 2010-02-06

- National Meteorological Center (January 1969). "Prospectus for an NMC Digital Facsimile Incoder Mapping Program" (PDF). Environmental Science Services Administration. Retrieved 2007-05-05

- Bridget R. Thomas; Elizabeth C. Kent & Val R. Swail (2005). "Methods to Homogenize Wind Speeds From Ships and Buoys" (PDF). International Journal of Climatology. John Wiley & Sons, Ltd. 25: 979–995. Bibcode:2005IJCli..25..979T. doi:10.1002/joc.1176. Retrieved 2011-01-29

- Stanislaw R. Massel (1996). Ocean surface waves: their physics and prediction. World Scientific. pp. 369–371. ISBN 978-981-02-2109-6. Retrieved 2011-01-18

- Hong Kong Observatory (2009-09-03). "The Hong Kong Observatory Computer System and Its Applications". The Government of the Hong Kong Special Administrative Region. Retrieved 2010-02-06

- Ronalds, B.F. (June 2016). "Sir Francis Ronalds and the Early Years of the Kew Observatory". Weather. doi:10.1002/wea.2739

- K. A. Browning; Robert J. Gurney (1999). Global energy and water cycles. Cambridge University Press. p. 62. ISBN 978-0-521-56057-3. Retrieved 2011-01-09

- Hydrometeorological Prediction Center (2000). "Hydrometeorological Prediction Center 1999 Accomplishment Report". National Oceanic and Atmospheric Administration. Retrieved 2007-05-05

- G. L. Timpe & N. Van de Voorde (October 1995). "NOMAD buoys: an overview of forty years of use". OCEANS '95. MTS/IEEE. Challenges of Our Changing Global Environment. Conference Proceedings. 1: 309–315. doi:10.1109/OCEANS.1995.526788

- William J. Emery; Richard E. Thomson (2001). Data analysis methods in physical oceanography. Gulf Professional Publishing. pp. 24–25. ISBN 978-0-444-50757-0. Retrieved 2011-01-09

- National Weather Service Forecast Office Honolulu, Hawaii (2010-02-07). "Pacific Streamline Analysis". Pacific Region Headquarters. Retrieved 2010-02-07

- Suomi, V. E.; Limaye, S. S.; Johnson, D. R. (1991). "High Winds of Neptune: A possible mechanism". Science. AAAS (USA). 251 (4996): 929–932. Bibcode:1991Sci...251..929S. PMID 17847386. doi:10.1126/science.251.4996.929

- Fahd, Toufic. : 842. Missing or empty |title= (help), in Rashed, Roshdi; Morelon, Régis (1996). Encyclopedia of the History of Arabic Science. 3. Routledge. pp. 813–852. ISBN 0-415-12410-7

An Overview of Atmospheric Thermodynamics

The thermodynamic phenomena that are behind the formation of clouds, changing in weather and temperature are known as atmospheric thermodynamics. Evaporation and condensation are two phenomena that provide the transfer of energy. This field is important for the modeling of particular regions and calamities like hurricanes and tornadoes. The topics discussed in the chapter are of great importance to broaden the existing knowledge on atmospheric thermodynamics.

Atmospheric Thermodynamics

Thermodynamics essentially deals with energy transformations in a system and its equilibrium states under such transformations. A system could be either open (no boundaries) or closed (boundaries). An *open system* exchanges energy and matter with its surrounding in attaining its equilibrium state, whereas a *closed system* has boundaries which are impermeable to matter but exchange of energy is possible as the system is not *isolated*. Thermodynamics plays a central role across a number of traditional disciplines of physics and chemistry because pressure and temperature are the key variables in the understanding the equilibrium states of a given system. The atmosphere and ocean are also complex systems where the exchanges of mass and energy are so intricate that their dynamics cannot be explained satisfactorily without appropriate consideration of thermodynamics. Thus, simulation of atmospheric phenomena ranging from cloud microphysics to large-scale atmospheric motions, involves the basic laws of dynamics and thermodynamics. The thermodynamic state of the atmosphere is greatly modified by the water vapour which on transformation of phase adds heat to the system. Also, evaporation of liquid water is accompanied with the cooling of the system. The thermodynamics of the moist air is thus different from dry air under certain transformations with phase changes. We first derive expressions for lapse rate and stability relevant for the dry atmosphere using the first law of thermodynamics assuming atmosphere in hydrostatic equilibrium. Finally, the thermodynamics of moist air will be discussed.

Lapse Rate

The lapse rate is the rate at which atmospheric temperature decreases with an increase in altitude. The terminology arises from the word *lapse* in the sense of a decrease or decline. While most often applied to Earth's troposphere, the concept can be extended to any gravitationally supported parcel of gas.

Definition

A formal definition from the *Glossary of Meteorology* is:

> The decrease of an atmospheric variable with height, the variable being temperature unless otherwise specified.

In general, a lapse rate is the negative of the rate of temperature change with altitude change, thus:

$$\gamma = -\frac{dT}{dz}$$

where γ is the lapse rate given in units of temperature divided by units of altitude, T = temperature, and z = altitude.

Convection and Adiabatic Expansion

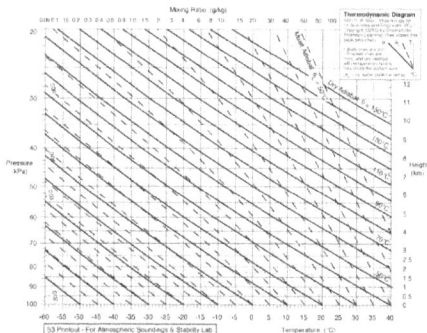

Emagram diagram showing variation of dry adiabats (bold lines) and moist adiabats (dash lines) according to pressure and temperature

The temperature profile of the atmosphere is a result of an interaction between radiation and convection. Sunlight hits the ground and heats it. The ground then heats the air at the surface. If radiation were the only way to transfer heat from the ground to space, the greenhouse effect of gases in the atmosphere would keep the ground at roughly 333 K (60 °C; 140 °F), and the temperature would decay exponentially with height.

However, when air is hot, it tends to expand, which lowers its density. Thus, hot air tends to rise and transfer heat upward. This is the process of convection. Convection comes to equilibrium when a parcel of air at a given altitude has the same density as the other air at the same elevation.

When a parcel of air expands, it pushes on the air around it, doing work (thermodynamics). Since the parcel does work but gains no heat, it loses internal energy so that its temperature decreases. The process of expanding and contracting without exchanging heat is an adiabatic process. The term *adiabatic* means that no heat transfer occurs into or out of the parcel. Air has low thermal conductivity, and the bodies of air involved are very large, so transfer of heat by conduction is negligibly small.

The adiabatic process for air has a characteristic temperature-pressure curve, so the process determines the lapse rate. When the air contains little water, this lapse rate is known as the dry adiabatic lapse rate: the rate of temperature decrease is 9.8 °C/km (5.38 °F per 1,000 ft) (3.0 °C/1,000 ft). The reverse occurs for a sinking parcel of air.

Note that only the troposphere (up to approximately 12 kilometres (39,000 ft) of altitude) in the Earth's atmosphere undergoes convection: the stratosphere does not generally convect. However, some exceptionally energetic convection processes -- notably volcanic eruption columns and overshooting tops associated with severe supercell thunderstorms -- may *locally* and *temporarily* inject convection through the tropopause and into the stratosphere.

The mathematics of the adiabatic lapse rate can be derived from thermodynamics, which defines an adiabatic process via:

$$PdV = -VdP / \gamma$$

the first law of thermodynamics can be written as

$$mc_v dT - Vdp / \gamma = 0$$

Also since : $\alpha = V / m$ and : $\gamma = c_p / c_v$ we can show that:

$$c_p dT - \alpha dP = 0$$

where c_p is the specific heat at constant pressure and α is the specific volume.

Assuming an atmosphere in hydrostatic equilibrium:

$$dP = -\rho g dz$$

where g is the standard gravity and ρ is the density. Combining these two equations to eliminate the pressure, one arrives at the result for the dry adiabatic lapse rate (DALR),

$$\Gamma_d = -\frac{dT}{dz} = \frac{g}{c_p} = 9.8 \,^\circ C / km$$

Moist Adiabatic Lapse Rate

The presence of water within the atmosphere complicates the process of convection. Water vapor contains latent heat of vaporization. As air rises and cools, it eventually becomes saturated and cannot hold its quantity of water vapor. The water vapor condenses (forming clouds), and releases heat, which changes the lapse rate from dry below the cloud to moist in the cloud. The release of latent heat is an important source of energy in the development of thunderstorms.

The moist adiabatic lapse rate varies strongly with temperature. A typical value is around 5 °C/km, (9 °F/km, 2.7 °F/1,000 ft, 1.5 °C/1,000 ft).

The saturated adiabatic lapse rate is given approximately by:

$$\Gamma_w = g \frac{1 + \dfrac{H_v r}{R_{sd} T}}{c_{pd} + \dfrac{H_v^2 r}{R_{sw} T^2}} = g \frac{R_{sd} T^2 + H_v r T}{c_{pd} R_{sd} T^2 + H_v^2 r \epsilon}$$

where:	
Γ_w	= Wet adiabatic lapse rate, K/m
g	= Earth's gravitational acceleration = 9.8076 m/s²
H_v	= Heat of vaporization of water, = 2501000 J/kg
R_{sd}	= Specific gas constant of dry air = 287 J kg⁻¹ K⁻¹
R_{sw}	= Specific gas constant of water vapour = 461.5 J kg⁻¹ K⁻¹
$\epsilon = \dfrac{R_{sd}}{R_{sw}}$	=The dimensionless ratio of the specific gas constant of dry air to the specific gas constant for water vapour = 0.622
e	= The water vapour pressure of the saturated air
p	= The pressure of the saturated air
$r = \epsilon e / (p - e)$	= The mixing ratio of the mass of water vapour to the mass of dry air
	= Temperature of the saturated air, K
c_{pd}	= The specific heat of dry air at constant pressure, = 1003.5 J kg⁻¹ K⁻¹

Environmental Lapse Rate

The environmental lapse rate (ELR), is the rate of decrease of temperature with altitude in the stationary atmosphere at a given time and location. As an average, the International Civil Aviation Organization (ICAO) defines an international standard atmosphere (ISA) with a temperature lapse rate of 6.49 K/km (3.56 °F or 1.98 °C/1,000 ft) from sea level to 11 km (36,090 ft or 6.8 mi). From 11 km up to 20 km (65,620 ft or 12.4 mi), the constant temperature is −56.5 °C (−69.7 °F), which is the lowest assumed temperature in the ISA. The standard atmosphere contains no moisture. Unlike the idealized ISA, the temperature of the actual atmosphere does not always fall at a uniform rate with height. For example, there can be an inversion layer in which the temperature increases with altitude.

Effect on Weather

The latent heat of vaporization adds energy to clouds and storms.

The varying environmental lapse rates throughout the Earth's atmosphere are of critical importance in meteorology, particularly within the troposphere. They are used to determine if the parcel of rising air will rise high enough for its water to condense to form clouds, and, having formed clouds, whether the air will continue to rise and form bigger shower clouds, and whether these clouds will get even bigger and form cumulonimbus clouds (thunder clouds).

As unsaturated air rises, its temperature drops at the dry adiabatic rate. The dew point also drops (as a result of decreasing air pressure) but much more slowly, typically about −2 °C per 1,000 m. If unsaturated air rises far enough, eventually its temperature will reach its dew point, and condensation will begin to form. This altitude is known as the lifting condensation level (LCL) when mechanical lift is present and the convective condensation level (CCL) when mechanical lift is absent, in which case, the parcel must be heated from below to its convective temperature. The cloud base will be somewhere within the layer bounded by these parameters.

The difference between the dry adiabatic lapse rate and the rate at which the dew point drops is around 8 °C per 1,000 m. Given a difference in temperature and dew point readings on the ground, one can easily find the LCL by multiplying the difference by 125 m/°C.

If the environmental lapse rate is less than the moist adiabatic lapse rate, the air is absolutely stable — rising air will cool faster than the surrounding air and lose buoyancy. This often happens in the early morning, when the air near the ground has cooled overnight. Cloud formation in stable air is unlikely.

If the environmental lapse rate is between the moist and dry adiabatic lapse rates, the air is conditionally unstable — an unsaturated parcel of air does not have sufficient buoyancy to rise to the LCL or CCL, and it is stable to weak vertical displacements in either direction. If the parcel is saturated it is unstable and will rise to the LCL or CCL, and either be halted due to an inversion layer of convective inhibition, or if lifting continues, deep, moist convection (DMC) may ensue, as a parcel rises to the level of free convection (LFC), after which it enters the free convective layer (FCL) and usually rises to the equilibrium level (EL).

If the environmental lapse rate is larger than the dry adiabatic lapse rate, it has a superadiabatic lapse rate, the air is absolutely unstable — a parcel of air will gain buoyancy as it rises both below and above the lifting condensation level or convective condensation level. This often happens in the afternoon mainly over land masses. In these conditions, the likelihood of cumulus clouds, showers or even thunderstorms is increased.

Meteorologists use radiosondes to measure the environmental lapse rate and compare it to the predicted adiabatic lapse rate to forecast the likelihood that air will rise. Charts of the environmental lapse rate are known as thermodynamic diagrams, examples of which include Skew-T log-P diagrams and tephigrams.

The difference in moist adiabatic lapse rate and the dry rate is the cause of foehn wind phenomenon (also known as "Chinook winds" in parts of North America). The phenomenon exists because warm moist air rises through orographic lifting up and over the top of a mountain range or large mountain. The temperature decreases with the dry adiabatic lapse rate, until it hits the dew point, where water vapor in the air begins to condense. Above that altitude, the adiabatic lapse rate decreases to the moist adiabatic lapse rate as the air continues to rise. Condensation is also commonly followed by precipitation on the top and windward sides of the mountain. As the air descends on the leeward side, it is warmed by adiabatic compression at the dry adiabatic lapse rate. Thus, the foehn wind at a certain altitude is warmer than the corresponding altitude on the windward side of the mountain range. In addition, because the air has lost much of its original water vapor content, the descending air creates an arid region on the leeward side of the mountain.

Adiabatic Lapse Rate

Consider the vertical movement of a parcel at pressure p, temperature T and of specific volume $\alpha(=1/\rho)$ in the atmosphere. Assume that atmosphere is in hydrostatic equilibrium; that is, *the upward force due to pressure gradient in vertical is balanced by the weight of the parcel.* The gravitational force and the buoyancy force are balanced; hence there is no need to consider them in the first law of thermodynamics. For a unit mass, the first law of thermodynamics states that a quantity dQ of heat added to the system is utilized inincreasing the internal energy of the system and as the work done on the system. Stated mathematically, the *First Law of Thermodynamics* is given as

$$dQ = dU + dW \qquad (1)$$

Internal energy increment: $dU = C_v dT$

Work done by pressure: $dW = p\,d\alpha$

$$dQ = C_v\,dT + p\,d\alpha \qquad (2)$$

C_v is the specific heat of air at constant volume. Differentiation of the equation of state gives

$$d(p\alpha) = d(R_d T) \Rightarrow p\,d\alpha + \alpha dp = R_d dT \qquad (3)$$

Using $\alpha = \dfrac{1}{\rho}$ and $R_d = C_p - C_v$ in (3), we can write it in the following form

$$pd\alpha + \frac{dp}{\rho} = R_d dT = \left(C_p - C_V\right)dT$$

Substituting the value of pdα in (2), we get

$$dQ = C_p dT - \frac{dP}{\rho} \Rightarrow dQ = C_p dT + gdz \qquad (4)$$

If the parcel undergoes adiabatic expansion or compression during its movement in the vertical, there is no exchange of heat (i.e.heat neither enters or leaves the parcel during motion); hence dQ = 0 and (4) gives

$$dQ = C_p dT + gdz = 0 \Rightarrow -\frac{dT}{dz} = \frac{g}{C_p} = \Gamma_d \qquad (5)$$

Equation gives the temperature decrease with height that can be calculated with *g= 9.81 ms⁻²* and C_p=*1005 J kg⁻¹* as

$$\Gamma_d = \frac{g}{C_p} = 9.76 \; K \; km^{-1} \qquad \text{...(6)}$$

This is the first important result in atmospheric thermodynamics. The adiabatic lapse rate Γ_d is the rate of change of temperature with height of a parcel of dry air when it adiabatically rises up or sinks down in a dry atmosphere. However, the profile of temperature from an ascent of a radiosonde gives the actual lapse rate $\Gamma = - dT/dz$ of the atmosphere. It varies widely due to the presence of water vapour and the average value of Γ is 6-7 K km^{-1} in the troposphere; that is $\Gamma < \Gamma_d$.

The meaning of constant lapse rate: If the atmosphere is heated by contact with the earth's surface and vertical motions set in, then heat will be distributed up in such a manner that the vertical temperature profile will have a constant gradient (Γ_d) of 10 K km^{-1} with altitude.

Now, consider the stability of the atmosphere with respect to vertical motions with the help of Figure that shows different vertical profiles of temperature (T vs z) in the atmosphere. In Figure a, different temperature profiles are given: profile marked OA is the dry adiabat (i.e. the temperature profile with constant lapse rate Γ_d); profile $CXYD$ is the actual temperature profile of the environment (actual lapse rate); and profile OB shows the lapse rate Γ greater than Γ_d, and dashed line profile shows that temperature of the atmosphere is uniform with height (isothermal layer) and $\Gamma = 0$. If a parcel initially at point X rises, then it will follow the dry adiabat line OA and would arrive at a location Q, where it is surrounded by ambient atmosphere with conditions at P on the profile CXYD. That is, parcel is warmer than the ambient atmosphere and shall continue to rise; hence the point X on profile CXYD is unstable. In other words, the parcels that are displaced from point X on the profile CXYD would never return to X. By similar argument point Y on profile CXYD is stable if it is displaced vertically; because parcels that are pushed up (down), being heavier (lighter) than the surrounding, shall return back to point Y. That is, atmosphere is stable above the point Y. Thus we have

(i) Atmosphere STABLE if $-\dfrac{\partial T}{\partial Z} < \Gamma_d$ ....(7)

(ii) Atmosphere UNSTABLE if $-\dfrac{\partial T}{\partial Z} > \Gamma_d$ ....(8)

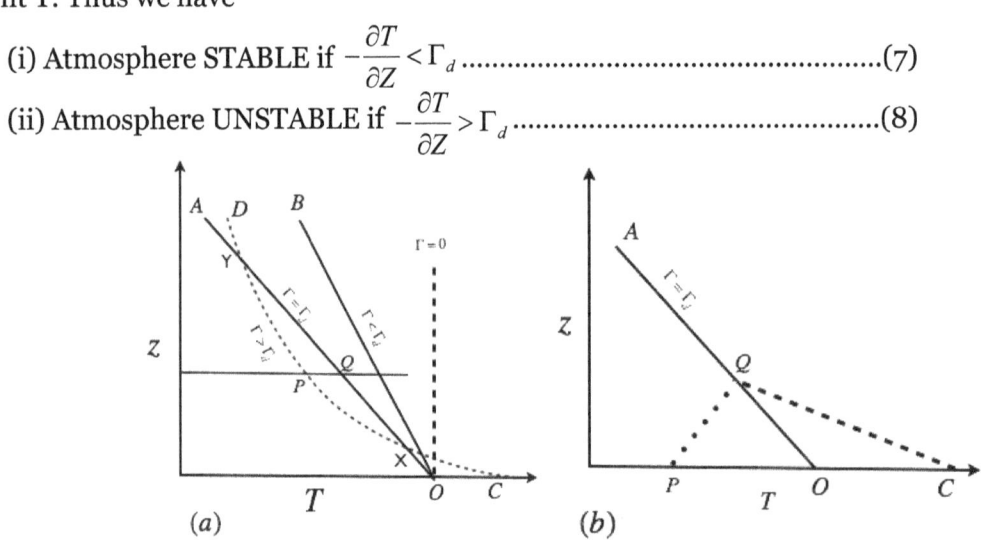

(a)   (b)

Atmospheric lapse rates: Panel (a):- The line marked OA is dry adiabat. The environmental temperature profile is the dotted line CXPYD. The part CX represents steep fall in temperature with height in the environment; Panel (b):- Profile OA corresponds to dry adiabatic lapse rate; the dotted line PQ shows rise in temperatures with height (inversion or negative lapse rate); the dashed line CQ corresponds to "superadiabatic lapse rate" in the atmosphere. Negative lapse rate corresponds to an adiabat sloping on the right as shown by the inclination of line PQ.

Potential Temperature

The potential temperature of a parcel of fluid at pressure P is the temperature that the parcel would acquire if adiabatically brought to a standard reference pressure P_0, usually 1000 millibars. The potential temperature is denoted θ and, for air, is often given by

$$\theta = T\left(\frac{P_0}{P}\right)^{R/c_p},$$

where T is the current absolute temperature (in K) of the parcel, R is the gas constant of air, and c_p is the specific heat capacity at a constant pressure. $R/c_p = 0.286$ for air (meteorology).

Contexts

The concept of potential temperature applies to any stratified fluid. It is most frequently used in the atmospheric sciences and oceanography. The reason that it is used in both fluids is that changes in pressure result in warmer fluid residing under colder fluid- examples being the fact that air temperature drops as one climbs a mountain and water temperature can increase with depth in very deep ocean trenches and within the ocean mixed layer. When potential temperature is used instead, these apparently unstable conditions vanish.

Comments

Potential temperature is a more dynamically important quantity than the actual temperature. This is because it is not affected by the physical lifting or sinking associated with flow over obstacles or large-scale atmospheric turbulence. A parcel of air moving over a small mountain will expand and cool as it ascends the slope, then compress and warm as it descends on the other side- but the potential temperature will not change in the absence of heating, cooling, evaporation, or condensation (processes that exclude these effects are referred to as dry adiabatic). Since parcels with the same potential temperature can be exchanged without work or heating being required, lines of constant potential temperature are natural flow pathways.

Under almost all circumstances, potential temperature increases upwards in the atmosphere, unlike actual temperature which may increase or decrease. Potential temperature is conserved for all dry adiabatic processes, and as such is an important quantity in the planetary boundary layer (which is often very close to being dry adiabatic).

Potential temperature is a useful measure of the static stability of the unsaturated atmosphere. Under normal, stably stratified conditions, the potential temperature increases with height,

$$\frac{\partial \theta}{\partial z} > 0$$

and vertical motions are suppressed. If the potential temperature decreases with height,

$$\frac{\partial \theta}{\partial z} < 0$$

the atmosphere is unstable to vertical motions, and convection is likely. Since convection acts to quickly mix the atmosphere and return to a stably stratified state, observations of decreasing potential temperature with height are uncommon, except while vigorous convection is underway or during periods of strong insolation. Situations in which the equivalent potential temperature decreases with height, indicating instability in saturated air, are much more common.

Since potential temperature is conserved under adiabatic or isentropic air motions, in steady, adiabatic flow lines or surfaces of constant potential temperature act as streamlines or flow surfaces, respectively. This fact is used in isentropic analysis, a form of synoptic analysis which allows visualization of air motions and in particular analysis of large-scale vertical motion.

Potential Temperature Perturbations

The atmospheric boundary layer (ABL) potential temperature perturbation is defined as the difference between the potential temperature of the ABL and the potential temperature of the free atmosphere above the ABL. This value is called the potential temperature deficit in the case of a katabatic flow, because the surface will always be colder than the free atmosphere and the PT perturbation will be negative.

Derivation

The enthalpy form of the first law of thermodynamics can be written as:

$$dh = T\,ds + v\,dp,$$

where dh denotes the enthalpy change, T the temperature, ds the change in entropy, v the specific volume, and P the pressure.

For adiabatic processes, the change in entropy is 0 and the 1st law simplifies to:

$$dh = v\,dp.$$

For approximately ideal gases, such as the dry air in the Earth's atmosphere, the equation of state, $pv = RT$ can be substituted into the 1st law yielding, after some rearrangement:

$$\frac{dp}{p} = \frac{c_p}{R}\frac{dT}{T},$$

where the $dh = c_p dT$ was used and both terms were divided by the product pv

Integrating yields:

$$\left(\frac{p_1}{p_0}\right)^{R/c_p} = \frac{T_1}{T_0},$$

and solving for T_0, the temperature a parcel would acquire if moved adiabatically to the pressure level p_0, you get:

$$T_0 = T_1 \left(\frac{p_0}{p_1} \right)^{R/c_p} \equiv \theta.$$

Potential Virtual Temperature

The potential virtual temperature θ_v, defined by

$$\theta_v = \theta \left(1 + 0.61r - r_L \right),$$

is the theoretical potential temperature of the dry air which would have the same density as the humid air at a standard pression P_0. It is used as a practical substitute for density in buoyancy calculations. In this definition θ is the potential temperature, r is the mixing ratio of water vapor, and r_L is the mixing ratio of liquid water in the air.

Related Quantities

The Brunt–Väisälä frequency is a closely related quantity that uses potential temperature and is used extensively in investigations of atmospheric stability.

The potential temperature θ of an air parcel is defined as the temperature that an air parcel would have if it were expanded or compressed adiabatically from its existing pressure (p) and temperature (T) to a standard pressure p_0 (generally taken as 1000 hPa).From the first law of Thermodynamics, we have the relation

$$dQ = C_p dT - \frac{dp}{\rho} \qquad \text{..(1)}$$

Eqn.(1) can be written for an adiabatic transformation (dQ=0) as

$$dQ = C_p dT - \alpha \, d \, p = 0 \Rightarrow \frac{C_p}{R_d} \frac{dT}{T} - \frac{dp}{p} = 0 \qquad \text{..(2)}$$

On integrating equation (2), we obtain

$$\frac{C_p}{R_d} \int_\theta^T \frac{dT}{T} = \int_{P_0}^p \frac{dp}{p} \qquad \left(\because \theta = T_0 \text{ at } p = p_0 \right) \Rightarrow \frac{C_p}{R_d} \ln\frac{T}{\theta} = \ln\frac{p}{p_0}$$

$$\theta = T \left(\frac{p_0}{p} \right)^{R_d/C_p} \qquad \text{..(3)}$$

The equation (3) is called the Poisson equation. Define

$$\kappa = \frac{R_d}{C_p} \text{ and } \gamma = \frac{C_p}{C_v} \text{ then } k = \left(1 - \frac{1}{\gamma} \right) = 0.2856$$

In calculating the value of κ, we have used

$$R_d = 287J\ K^{-1}kg^{-1}\ and\ C_p = 1005J\ K^{-1}kg^{-1}$$

Exercise 3: Starting with eq. (3), show that $\dfrac{T^{\gamma}}{p^{\gamma-1}} = \dfrac{T^{\gamma}_0}{p_0^{\gamma-1}} = \dfrac{\theta^{\gamma}}{p_0^{\gamma-1}} = const.$

Thus, a new temperature θ can be defined with (3) which a parcel will have if brought adiabatically to a reference pressure p_0, taken as 1000 hPa. It must be noted that for an incompressible medium, temperature is the appropriate variable; but for the compressible atmosphere, it is the potential temperature θ. To study the dynamic evolution of the atmospheric flows θ is an appropriate variable, because it is a measure of the sum of potential and internal energy and it is conserved in adiabatic motions of the atmosphere. Quantities that remain invariant during any transformation are said to be conserved variables. The potential temperature is therefore an extremely important parameter in atmospheric thermodynamics. The conservation of θ simplifies the treatment of dynamics of compressible fluids in the absence of heat sources and sinks, much like the dynamics of incompressible fluids that is rendered simpler (divergence free motions) because density remains constant. The conservation of θ is mathematically expressed as

$$\frac{d\theta}{dt} = 0 \qquad \dots\dots\dots\dots\dots\dots\dots\dots\dots\dots\dots\dots\dots\dots\dots\dots\dots\dots\dots\dots\dots\dots\dots(4)$$

Exercise 3: A parcel of air has a temperature -51°C at 250 hPa level. What is its potential temperature. What temperature will the parcel have if it were brought into the cabin of a jet aircraft and compressed adiabatically to a cabin pressure of 850 hPa? [Hint: Start with T = (-51 + 273) K, p = 250 hPa po = 1000 hPa; calculate θ. Parcel brought to cabin: T = 222 K, p = 250 hPa,po = 850 hPa; calculate θ]

Note: Taking the logarithm of (3), it becomes easier to differentiate (3); we obtain,

$$ln\theta = lnT + \frac{R}{C_P}lnP_0 - \frac{R}{C_P}ln\ p$$ Now differentiate on both sides to get

$$\frac{1}{\theta}\frac{d\theta}{dz} = \frac{1}{T}\frac{dT}{dz} - \frac{R}{C_P}\frac{1}{p}\frac{dp}{dz} \qquad \dots\dots\dots\dots\dots\dots\dots\dots\dots\dots\dots\dots\dots\dots\dots\dots(5)$$

Static Stability

In equation (5), if hydrostatic equation is used then we have,

$$\frac{1}{\theta}\frac{d\theta}{dz} = \frac{1}{T}\frac{dT}{dz} - \frac{R}{C_P}\frac{1}{p}(-g\rho)\ : \qquad \rho = \frac{p}{RT}$$

$$\frac{T}{\theta}\frac{d\theta}{dz} = \frac{dt}{dz} + \frac{g}{C_P} \qquad \dots\dots\dots\dots\dots\dots\dots\dots\dots\dots\dots\dots\dots\dots\dots\dots\dots\dots\dots\dots\dots(6)$$

For an atmosphere with constant θ (i.e. an adiabatic atmosphere), the atmospheric lapse rate -dT/

dz is obtained from (6) by setting $d\theta/dz=0$, and we obtained the earlier result,

$$-\frac{dt}{dz}=\frac{g}{C_p}=\Gamma_d$$ ....(7)

Hence, if potential temperature is a function of height, the atmospheric lapse rate $\Gamma = -dT/dz$ will differ from adiabatic lapse rate Γ_d and we have from (6),

$$\frac{T}{\theta}\frac{d\theta}{dz}=\Gamma_d-\Gamma$$ ....(8)

If $\Gamma < \Gamma_d$, it implies that the vertical gradient of θ is positive i.e. θ increase with height. Thus, an air parcel that undergoes an adiabatic displacement from its equilibrium position will be positively buoyant when displaced vertically downward. The air parcel will be negatively buoyant (sinks) when displaced vertically upward from its equilibrium position. Such as atmosphere is said to be statically stable or stably stratified atmosphere

Buoyancy Oscillations

Adiabatic oscillations of a fluid parcel about its equilibrium position in a stably stratified atmosphere are called buoyancy oscillations. The frequency of such oscillations can be derived by vertically displacing parcel upward or downward a small distance δz from its equilibrium position at a height z. The atmosphere is always considered in hydrostatic equilibrium. So any imbalance between the upward pressure gradient and its weight will result in vertical acceleration; but when there is a balalnce, we have

$$\frac{dp_0}{dz}=-g\rho_0 \Rightarrow -\frac{1}{\rho_0}\frac{\partial p_0}{\partial z}-g=0$$ ....(9)

Here p_0 and ρ_0 are pressure and density of the environment. If p and ρ are the pressure and density of the parcel, any imbalance between the vertically acting pressure gradient force and the weight of the parcel will result in vertical accelerations (dw/dt) of the parcel. The Newton's second law gives the equation of motion of the parcel as,

$$\frac{dw}{dt}=-\frac{1}{\rho}\frac{dp}{dz}-g \;\; or \;\; \rho\frac{dw}{dt}=-\frac{dp}{dz}-g\rho \;\; (parcel) \; and \; w=\frac{d}{dt}(\delta z) \; ; \;\; hence$$

$$\rho\frac{d^2(\delta z)}{dt^2}=-\frac{dp}{dz}-g\rho.$$ ....(10)

Since environment is in hydrostatic equilibrium, dp/dz in (10) is replaced by dp_0/dz and we get

$$\rho\frac{d^2(\delta z)}{dt^2}=-\frac{dp_0}{dz}-g\rho$$

Now replace dp_0/dz by $-g\rho_0$ in the above equation using (9), and (10) becomes,

$$\rho \frac{d^2(\delta z)}{dt^2} = g(\rho_0 - \rho)$$ ....(11)

Note that the term $g(\rho_0 - \rho)$ on the right hand side of (11) is the buoyancy force acting on the parcel of unit volume. We can now write (11) as

$$\frac{d^2}{dt^2}(\delta z) = g \frac{\rho_0 - \rho}{\rho}$$ ....(12)

As pointed out earlier that a parcel adjusts its pressure instantaneously with the surrounding pressure. Thus, for a parcel undergoing small displacements δz from its equilibrium position without disturbing its environment, we set p(parcel) = p_0(environ). Hence, the pressure gradient term in (10) is replaced by dp_0/dz in parcel method.

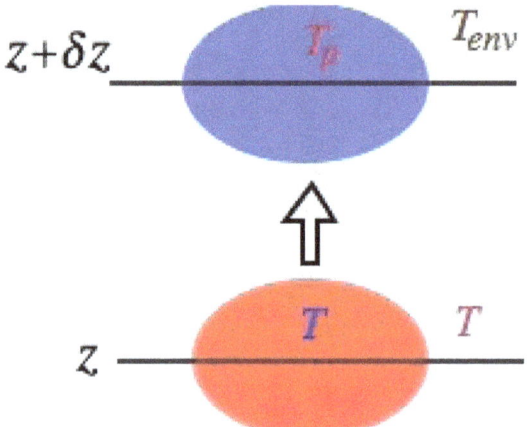

Displacement of a parcel

Consider a parcel in Figure at a height z and it is displaced vertically by δz to a new position $z+\delta z$. As the parcel moves from its equilibrium position z to new position $z+\delta z$, we can estimate the change in the thermodynamic variables T, θ and ρ. Initially, the parcel and environment are at the same temperature T, density and pressure, but when displaced its temperature is given by

$$T_p(z+\delta z) = T_p(z) + \frac{dT}{dz}\delta z \qquad or \; T_p(z+\delta z) = T_p(z) - \Gamma_d \delta z$$ ....(13a)

For the environment

$$T_{env}(z+\delta z) = T(z) + \frac{dT}{dz}\delta z \qquad or \; T_{env}(z+\delta z) = T(z) - \Gamma \delta z$$ ....(13b)

The density ρ of parcel and p_0 of environment can also be written at new position z_1. Now, one may derive the following expression that holds also at the new position of the parcel,

$$g \frac{\rho_0 - \rho}{\rho} = g \frac{T_p - T_{env}}{T_{env}}$$ ....(14)

Using (13) and (14) in (12), we get

$$\frac{d^2}{dt^2}(\delta z) = g\frac{(\Gamma - \Gamma_d)\delta z}{(T - \Gamma\delta z)} = g\frac{(\Gamma - \Gamma_d)\delta z}{T}(1 - \frac{\Gamma}{T}\delta z)^{-1} = g\frac{(\Gamma - \Gamma_d)\delta z}{T}\left(1 + \frac{\Gamma}{T}\delta z + ...\right)$$

$$\frac{d^2}{dt^2}(\delta z) = -N^2\delta z \ ; \ N^2 = \frac{g}{T}(\Gamma_d - \Gamma).$$ ....(15)

In eqn. (15), N^2 is the buoyancy frequency also called the Brunt-Väisälä frequency; N^2 is a measure of static stability of the atmosphere; for the lower atmosphere, the corresponding period $\tau = 2\pi/N$ is a few minutes.The parcel undergoes oscillations when $N^2 > 0$.

Exercise 4: If θ and θ_0 are the potential temperatures of the parcel and the environment, using (15), show that $g\frac{\rho_0 - \rho}{\rho} = g\frac{\theta - \theta_0}{\theta_0}$. The parcel undergoes adiabatic displacements, hence $\theta(z + \delta z) = \theta_0(z)$
Write for the environment, $\theta_0(z + \delta z) = \theta_0(z)\frac{d\theta_0}{dz}\delta z$ and show that $N^2 = \frac{g}{\theta_0}\frac{d\theta_0}{dz} = g\frac{d(ln\theta_0)}{dz}$
The solution of (15) is given by

$$\delta z = Ae^{iNt}$$ ....(16)

If parcel oscillates about its mean position when, $N^2 > 0$ then for average tropospheric conditions,

$$N = 1.2 \times 10^{-2}S^{-1} \quad \Rightarrow \quad \tau = \frac{2\pi}{N} = 8.7\,min$$

If $N = 0$, There will be no oscillations of the displaced parcel about its equilibrium position and parcel will be in *neutral equilibrium* (i.e. no accelerating force on parcel).

If $N^2 < 0$, the potential temperature will decrease with height; which means the displacement of the parcel will increase exponential with time; in other words, parcels continue to move through the atmospheric column and the column is unstable. The above condition also lead us to write the conditions of static stability of the atmosphere in terms of the potential temperature from the relation $N^2 = \frac{g}{\theta_0}\frac{d\theta_0}{dz}$ .Thus, we have,

$$
\begin{array}{lll}
N^2 > 0 & \textit{Atmosphere statically stable} & \dfrac{d\theta_0}{dz} > 0 \\[2mm]
N^2 = 0 & \textit{Statically neutral} & \dfrac{d\theta_0}{dz} = 0 \\[2mm]
N^2 < 0 & \textit{Statically unstable} & \dfrac{d\theta_0}{dz} < 0
\end{array}
$$ ....(17)

An important note: On the synoptic scale, atmosphere is always stably stratified and when the stratification is unstable, convective overturning removes quickly the unstable regions that develop. The convective overturning essentially mixes the air in overlaying layers and it redefines the distribution of isentropes (surfaces of equal θ) in the stratification. However, for moist atmosphere, situation is more complicated. If the temperature actually increases with height (temperature inversion) the atmosphere is gravitationally stable as warm air overlays the cold air. Temperature inversions inhibit mixing; and water vapour, aerosols and pollutants are trapped in the layer. During winter, temperature inversions occur during nights, therefore in the big cities, vehicular emissions will be trapped in this layer and there will be episodes of high pollution and smog after sunrise. However, it is worth mentioning that over global deserts, in day time hours, negative stability can exist in the air next to the ground with super-adiabatic lapse rates.

General Meteorological Problem

D. Brunt (1941) in his book Physical and Dynamical Meteorology has stated, "The main problem to be discussed in connection to the thermodynamics of the moist air is the variation of temperature produced by changes of pressure, which in the atmosphere are associated with vertical motion. When damp air ascends, it must eventually attain saturation, and further ascent produces condensation, at first in the form of water drops, and as snow in the later stages". This statement of the problem emphasizes the role of vertical ascends in producing condensation of water vapour. However, several text books and papers discuss this problem on the assumption that products of condensation are carried with the ascending air current and the process is strictly reversible; meaning that if the damp air and water drops or snow are again brought downwards, the evaporation of water drops or snow uses up the same amount of latent heat as it was liberated by condensation on the upward path of the air. Another assumption is that the drops fall out as the damp air ascends but then the process is not reversible, and Von Bezold (1883) termed it as a pseudo-adiabatic process. It must be pointed out that if the products of condensation are retained in the ascending current, the mathematical treatment is easier in comparison to the pseudo-adiabatic case. There are four stages that can be discussed in connection to the ascent of moist air.

(a) The air is saturated;

(b) The air is saturated and contains water drops at a temperature above the freezing-point

(d) Saturated air and ice at temperatures below 0°C.

However, clouds may be composed of supercooled water drops at temperatures as low as -40°C; it also forms an important stage in conjunction with the aforementioned four stages of moist air. To treat water vapour in the atmosphere, its mass mixed with a certain amount of air in a given volume must be known. The moisture content of air can be expressed in several ways that relate the partial pressure of water vapour to its volume or mass mixing ratio in the atmosphere.

Moisture Parameters

We define here the basic humidity variables, which essentially give an estimate of the amount of water in vapour form present in a given volume of moist air. These definitions could also be extended to other forms of hydrometeors which are at atmospheric pressure p and temperature T. The following definitions are used in for the moist air parcels.

(a) Mixing ratio (r): It is defined as the ratio of the mass of water vapour in a certain volume of air to the mass of dry air in that volume, so

$$r = \frac{m_v}{m_d} = \frac{\rho_v}{\rho_d};$$

$m_v = mass\ of\ water\ vapour$

$m_d = mass\ of\ dry\ air\ (kg\ /\ kg)$

(b) Specific humidity: It is defined as the mass of water vapour in a given mass of air which is a mixture of air and water vapour; thus,

$$q = \frac{m_v}{m_v + m_d} = \frac{r}{1+r}$$

Since r is only a few percent, the numerical values of r and q are nearly equal. Further, ideal gas law can be applied to each constituent of any mixture of gases, we may write the gas law for water vapour as

$$e = \rho_v R_v T$$

where e and ρ_v are the pressure and density of the water vapour and Rv is the gas constant for 1 kg of water vapour. The gas constant for water vapour R_v is calculated from the universal gas constant and molar mass water vapour as

$$R_v = 1000 \frac{R^*}{M_{H_2O}} = 1000 \frac{8.3145}{18.016} = 461.51 kg^{-1} K^{-1}; \quad R_d = 287 J\ kg^{-1} K^{-1}$$

The ratio of gas constants for dry air and water vapour is calculated as

$$= \frac{R_d}{R_v} = \frac{M_v}{M_d} = \frac{18.016}{28.97} = 0.622.$$

From Dalton's law of partial pressures of a mixture of gases, the moisture content of the air can also be defined in terms of its partial pressure in the atmosphere.

(c) Volume mixing ratio (r_v):

$$r_v = e / p$$

(d) Absolute Humidity: The concentration ρ_v of water vapour in air is called the absolute humidity (g/m³). It can be easily obtained from the ideal gas law of water vapour.

Example 1: At 0°C, = 1.275 kg/m3, and = 4.770 x 10⁻³ kg m⁻³. Find the total pressure exerted by the mixture.

Solution:

$$p_d = (1.275 \frac{kg}{m^3})(287.0\ JK^{-1} kg^{-1})(273.2K) \Rightarrow P_d = 999970.71 Pa$$

$$e = \rho_v R_v T = \left(4.770 \times 10^{-3} \frac{kg}{m^3}\right)\left(461.51 \frac{J}{K\ kg}\right)(273.2K) = 601.4232 Pa$$

Hence total pressure of the mixture p = 999.71 + 6.014 = 1005.72 hPa

Virtual Temperature

In atmospheric thermodynamics, the virtual temperature T_v of a moist air parcel is the temperature at which a theoretical dry air parcel would have a total pressure and density equal to the moist parcel of air.

Introduction

Description

In atmospheric thermodynamic processes, it is often useful to assume air parcels behave approximately adiabatic, and thus approximately ideally. The specific gas constant for the standardized mass of one kilogram of a particular gas is variable, and described mathematically as:

$$R_x = \frac{R^*}{M_x},$$

where R^* is the molar gas constant and M_x is the apparent molar mass of gas x in kilograms per mole. The apparent molar mass of a theoretical moist parcel in Earth's atmosphere can be defined in components of water vapor and dry air as:

$$M_{air} = \frac{e}{p}M_v + \frac{p_d}{p}M_d,$$

with e partial pressure of water, p_d dry air pressure, and M_v and M_d representing the molar masses of water vapor and dry air respectively. The total pressure p is described by Dalton's Law of Partial Pressures:

$$p = p_d + e.$$

Purpose

Rather than carry out these calculations, it is convenient to scale another quantity within the ideal gas law to equate the pressure and density of a dry parcel to a moist parcel. The only variable quantity of the ideal gas law independent of density and pressure is temperature. This scaled quantity is known as virtual temperature, and it allows for the use of the dry-air equation of state for moist air. Temperature has an inverse proportionality to density. Thus, analytically, a higher vapor pressure would yield a lower density, which should yield a higher virtual temperature in turn.

Derivation

Consider a moist air parcel containing masses m_d and m_v of dry air and water vapor in a given volume V. The density is given by:

$$\rho = \frac{m_d + m_v}{V} = \rho_d + \rho_v,$$

where ρ_d and ρ_v are the densities the dry air and water vapor would respectively have when occupying the volume of the air parcel. Rearranging the standard ideal gas equation with these variables gives:

$e = \rho_v R_v T$ and $p_d = \rho_d R_d T$.

Solving for the densities in each equation and combining with the law of partial pressures yields:

$$\rho = \frac{p-e}{R_d T} + \frac{e}{R_v T}.$$

Then, solving for p and using $\epsilon = \dfrac{R_d}{R_v} = \dfrac{M_v}{M_d}$ is approximately 0.622 in Earth's atmosphere:

$$p = \rho R_d T_v,$$

where the virtual temperature T_v is:

$$T_v = \frac{T}{1 - \dfrac{e}{p}(1-\epsilon)}.$$

We now have a non-linear scalar for temperature dependent purely on the unitless value e/p, allowing for varying amounts of water vapor in an air parcel. This virtual temperature T_v in units of Kelvin can be used seamlessly in any thermodynamic equation necessitating it.

Variations

Often the more easily accessible atmospheric parameter is the mixing ratio w. Through expansion upon the definition of vapor pressure in the law of partial pressures as presented above and the definition of mixing ratio:

$$\frac{e}{p} = \frac{w}{w+\epsilon},$$

which allows:

$$T_v = T\frac{w+\epsilon}{\epsilon(1+w)}.$$

Algebraic expansion of that equation, ignoring higher orders of w due to its typical order in Earth's atmosphere of 10^{-3}, and substituting ϵ with its constant value yields the linear approximation:
$T_v \approx T(1+0.61w)$.

An approximate conversion using T in degrees Celsius and mixing ratio w in g/kg is:

$$T_v \approx T + \frac{w}{6}.$$

Uses

Virtual temperature is used in adjusting CAPE soundings for assessing available convective potential energy from Skew-T log-P diagrams. The errors associated with ignoring virtual temperature correction for smaller CAPE values can be quite significant. Thus, in the early stages of convective storm formation, a virtual temperature correction is significant in identifying the potential intensity in tropical cyclogenesis.

In meteorology, it is customary to define a virtual temperature to take in to consideration effect of water vapour in the atmosphere. Since $M_v < M_d$ and $R_v > R_d$, therefore instead of using the gas constant R_v, it is convenient to use R_d and fictitious temperature T_v (because never measured) in the equation of state for water vapour. Thus,

$$p = \rho R_d T_v \quad \text{..(1)}$$

Here T_v is the virtual temperature; its expression is now derived starting with the density of mixture , $\rho = \rho_d + \rho_v$; $\rho_{d=}\rho_d R_d T$ $\;and\;\; e = \rho_v R_v T$ The pressure of the mixture is given as,

$$p = p_d + e \quad or \quad p_d = p - e$$

Since $\rho_d = \dfrac{p-e}{R_d T}$; $\rho_v = \dfrac{e}{R_v T}$ we can write the density of the mixture ρ as

$$\rho = \rho_d + \rho_v = \frac{p-e}{R_d T} + \frac{e}{R_v T} = \frac{p}{R_d T} + e\left[\frac{1}{R_v T} - \frac{1}{R_d T}\right] \quad \text{which can be further written as,}$$

$$\rho = \frac{p}{R_d T}\left[1 + \frac{e}{p}\left(\frac{R_d}{R_v} - 1\right)\right] \quad or \quad \rho = \frac{p}{R_d T}\left[1 - \frac{e}{p}(1-\varepsilon)\right] \quad \text{...(2)}$$

Now (1) can be rearranged as

$$p = \rho R_d \frac{T}{1 - \dfrac{e}{p}(1-\varepsilon)} \quad and \quad \varepsilon = \frac{R_d}{R_v} = \frac{287}{461.51} = 0.622$$

$$\text{...(3)}$$

If we define $T_v = \dfrac{T}{1 - \dfrac{e}{p}(1-\varepsilon)}$ then the pressure p in (3) can be written as given in the equation above

with and p as density and pressure of the air respectively.

Thus for the moist air, actual temperature is replaced by virtual temperature in the equation of state; it is in fact used as a state variable in moist air dynamics. Also one may define the virtual temperature, as that temperature the dry air would attain so as to have the same density as that of the moist air at the same pressure.

Example 2: Given a total pressure p =1026.8 hPa, find the vapour pressure e in a mixture of water vapour and air if the water vapour mixing ratio r = 5.5 g/kg.

Solution:

$$r = \frac{\rho_v}{\rho_d} = \left(\frac{e}{R_v T}\right)\left(\frac{R_d T}{p-e}\right) = \frac{e}{p-e}\cdot\frac{R_d}{R_v} = \frac{\varepsilon e}{p-e}. \quad Hence \quad e = \frac{r}{r+\varepsilon}p$$

Put r=5.5 g/kg and p = 1026.8 hPa = 102680 Pa in the above expression of e, and we get,

$$e = \frac{(5.5 \times 10^{-3})(102680)}{0.622 + 5.5 \times 10^{-3}} = 900.70 \; Pa \;\; \text{or water vapour pressure e = 9.00 hPa.}$$

Exercise 1: Starting from the definition of virtual temperature, show that $T_v = (1 + 0.61q)T$. *Also find the expression for q.*

Relation between Moisture Variables and Water Vapour Pressure

Humidity variables can also be defined in terms of water vapour pressure. The relation between mass mixing ratio r and e is derived easily in the following manner. Starting with the definition of mixing ratio r given in, we have

$$r = \frac{\rho_v}{\rho_d} = \left(\frac{e}{R_v T}\right)\left(\frac{R_d T}{p - e}\right) = \frac{\varepsilon e}{p - e} \qquad\qquad\qquad (3)$$

The specific humidity is related to the mass mixing ratio r as

$$q = \frac{\rho_v}{\rho} = \frac{\rho_v}{\rho_d + \rho_v} = \frac{r}{1 + r}, \text{ as defined in (4.2). Now use the relation (4.8) for } r \text{ to get } q \text{ as,}$$

$$q = \left(\frac{\varepsilon e}{p - e}\right)\left(1 + \frac{\varepsilon e}{p - e}\right)^{-1} = \frac{\varepsilon e}{p - (1 - \varepsilon)e} \qquad\qquad\qquad (4)$$

Saturated and Unsaturated Air

The most commonly used phrases for moist air are indeed misleading. Phrases such as *"air is saturated with water vapour"; or "air can hold no more water vapour"; and "warm air can hold more moisture than cold air"* are quite commonly used in weather descriptions. All these phrases suggest that air absorbs moisture much like a sponge, which is not the case with air. According to Dalton's principle of partial pressures for a mixture of gases, each component exerts a partial pressure and sum of the partial pressures is the total pressure exerted by the mixture. Hence, the exchange of water molecules between the liquid and vapour phases is indeed independent of the presence of other components in air. Moreover, an equilibrium vapour pressure is established between the liquid and vapour phases when a certain number of water molecules leave the liquid phase at a given temperature and an equal number of water molecules from vapour phase enter the liquid phase. Air is said to be saturated at such equilibrium, otherwise it is unsaturated. Both terms are used for characterizing dry air and moist air.

Unsaturated air: When liquid and vapour phase are imagined to coexist and the rate of evaporation of water from liquid exceed rate of condensation, then air is unsaturated at a given temperature T and it can be expressed by the pair (e,T) .

Saturated air: When the vapour pressure in the gas phase reach to a value (say e_s) such that the rate of evaporation of water from liquid is equal to rate of condensation of vapour in air at temperature T of the system. The pair (e_s,T) may express the state of the system. Normally the interface of liquid and vapour phase is regarded as a plane surface.

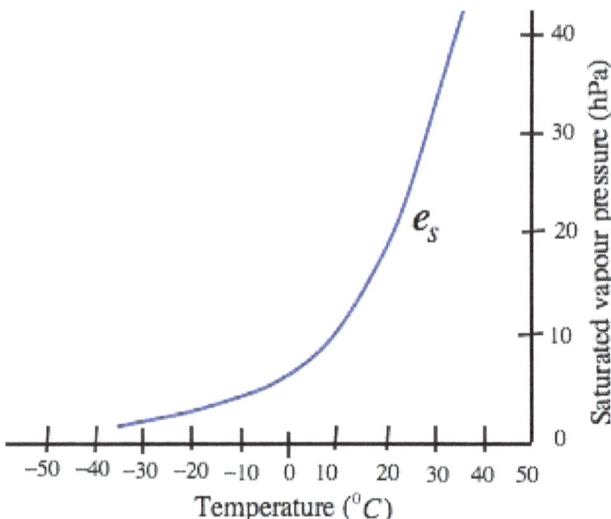

Variation of saturation vapour pressure e_s over plane surface of pure water with temperature.

Saturated mixing ratio: It is the mass of a given volume of air that is saturated with respect to plane surface of pure water, to the mass of dry air in a given volume; that is,

$$r_s = \frac{m_{vs}}{m_d} = \frac{\rho_{vs}}{\rho_d} = \left(\frac{e_s}{R_v T}\right)\left(\frac{p-e_s}{R_d T}\right)^{-1} = 0.622\frac{e_s}{p-e_s} = \frac{\varepsilon e_s}{p-e_s} \qquad ....(5)$$

Because $p >> e_s$, the eqn. (5) can be put in the following simplified form

$$r_s \approx 0.622\frac{e_s}{p} = \frac{\varepsilon e_s}{p} \qquad ....(6)$$

That is, at a given temperature, r_s is inversely proportion to p. Note that $e_s = e_s(T)$, so the saturated mixing ratio r_s is also a function of temperature; this implies that r_s increases with increasing temperature at constant pressure. But at constant temperature, r_s increases with decreasing pressure p, which is evident from (4.11). The curve $e_s = e_s(T)$ is shown in Figure as a function of temperature; e_s decreases with height in the troposphere. Since atmospheric pressure also decreases with altitude, that is why the rising parcels of unsaturated air turn saturated leading to cloud formation and rains. The process is hastened in the troposphere because temperature also decreases with altitude.

Relative humidity: The relative humidity r_h is the ratio of mass of water vapour in a sample of moist air of volume V, to the mass of water vapour if the air in volume V were saturated. Thus, we have

$$r_h = \frac{m_v}{m_{vs}} = \frac{r}{r_s} = \frac{e}{e_s} \qquad ....(7)$$

The reported relative humidity is generally expressed in per cent; i.e. RH= $r_h \times 100$.

Isobaric cooling: Imagine that temperature of a parcel of moist air reduces continuously but its pressure remains the same. During this process, e will not change but e_s will decrease as the temperature falls during cooling. As a consequence, under isobaric cooling, the relative humidity of the parcel will increase with decreasing temperature. If temperature continues to cool further down, then phase change of water vapour to liquid or ice may happen when threshold temperature is reached.

Dew point temperature (T_d): It is that temperature to which air must be cooled at constant pressure (i.e. isobarically) for it to become saturated with respect to plane surface of pure water. Thus, we have $r(T, p) = r_s(T_d, p)$ and the relative humidity at temperature T and pressure p is given by

$$RH = \frac{r_s(T_d, p)}{r_s(T, P)} \times 100.$$

A Thumb Rule: For moist air (RH > 50%), T_d decreases by 1°C for every 5% decrease in RH. Thus, starting at $T_d = T$ (dry bulb temperature) where RH = 85%, then

$$T_d = T - \left(\frac{100 - 85}{5}\right) = T - 3^0 C \ or \ T - T_d = 3^0 C$$

Frost-Point Temperature: It is that temperature to which air must be cooled isobarically for it to become saturated with respect to plane surface of ice.

Example 3: What are RH and T_d for air at 1000 hPa and 18°C having a mixing ratio 6 g kg⁻¹ Saturation mixing ratio is 13 g kg⁻¹ at this state.

Solution:

RH = 46% from the thumb rule T_d=7.2 °C.

However, to find the actual value, we use the Tephigram. One has to move from right to left along 1000 hPa line to intercept the saturation line of magnitude 6 g/kg; this occurs at 6.5°C; this is the dew point temperature in this case.

On the globe, dew point is a good indicator of moisture levels in the air as the pressure typically varies only by a few per cent spatially and temporally. Over warm bodies, highest dew point temperatures are noted, though complete saturation is absent in hot area.

Thermodynamics Unsaturated Moist Air

It has been shown that the equation of state for dry air can be applied to moist air when T is replaced by T_v, that is

$$p = \rho R_d T_V \qquad\qquad \rho = \rho_d + \rho_v$$
$$p = \rho R_m T \qquad R_m = (1 + 0.61q) R_d$$

These two expressions are equivalent; here R_m is the gas constant for moist air. Since moist air is a mixture of dry air and water vapour, *the specific heat at constant volume of the mixture is determine by considering one kilogram of dry air and ξ kilogram of water vapour.* Next, consider adding heat dq to the sample which raises its temperature by dT; we thus have

$$(1+\xi)dq = (C_v)_{air}\, dT + \xi(C_v)_{vap}\, dT; \quad volume(V) = const.$$

Because , $(C_v)_{mix} = \left(\dfrac{\partial q}{\partial T}\right)_v$ we have

$$(1+\xi)(C_v)_{mix} = (C_v)_{air} + \xi(C_v)_{vap}$$

$$(C_v)_{mix} = \left(\frac{\partial q}{\partial T}\right)_v = \frac{(C_v)_{air} + \xi(C_v)_{vap}}{1+\xi}; \quad V = const.$$

Since $(C_v)_{air}$ at STP = 718 $JK^{-1}kg^{-1}$, $(C_v)_{vap}$ at STP=1390 $JK^{-1}kg^{-1}$, eqn. (4.15) states that $(C_v)_{mix}$ in a parcel is the mass weighted mean of $(C_v)_{air}$ and $(C_v)_{vap}$. In the same manner, we can calculate $(C_p)_{mix}$ as a mass-weighted mean of $(C_p)_{air}$ and $(C_p)_{vap}$ as

$$C_{pm} = (C_p)_{mix} = \frac{(C_p)_{air} + \xi(C_p)_{vap}}{1+\xi}$$

Since, $(C_p)_{air}$ at STP = 1005 $JK^{-1}kg^{-1}$; $(C_p)_{vap}$= 1850 $JK^{-1}kg^{-1}$ the specific heat of the mixture with pressure remaining constant can be calculated from relation. Knowing the specific heat of the mixture at constant pressure, so long as there is no condensation of water vapour during the lifting of parcels, we can calculate the lapse rate of unsaturated moist air parcel as

$$\Gamma_m = \frac{g}{(C_p)_{mix}} = \frac{g}{C_{pm}}; \quad \Gamma_m < \Gamma_d$$

Lifting Condensation Level (LCL)

The lifting condensation level (LCL) is defined as the level to which an *unsaturated* parcel of moist air should be *lifted adiabatically* in order to become *saturated*. The process is explained in the Figure. During lifting, the mixing ratio r and the potential temperature θ remain constant but r_s decreases until it becomes equal to r at the LCL. Hence the lifting condensation levelis the point C on Figure, to which a parcel has been lifted along the line AC from its initial position at a point A(p, T, r) with pressure p, temperature T and mixing ratio r. It is required to locate the point C on a *Tephigram* for issuing the local forecast. Mathematically, one can find the LCL and the corresponding Z_{LCL}, T_{LCL} and P_{LCL} can be calculated from thermodynamic relations using the condition that the saturation mixing ratio of a parcel at point C is equal to its mixing ratio at A. Note that point C in Figure is the intersection of two lines: *q =const. and* r_s *= const.* Observations provide values of p, T, and r at point A. The saturation mixing ratio line is drawn with r_s =const. In order to locate the point B, the parcel is cooled isobarically; that is, the parcel is moved horizontally to arrive at intersection point B, where its temperature is the dew point temperature T_d, and air parcel at B is saturated. From point B, move along the line r_s =const. up to a point C where it intersect the dry adiabat through the point A. The LCL is also referred to as the adiabatic condensation temperature in the literature.

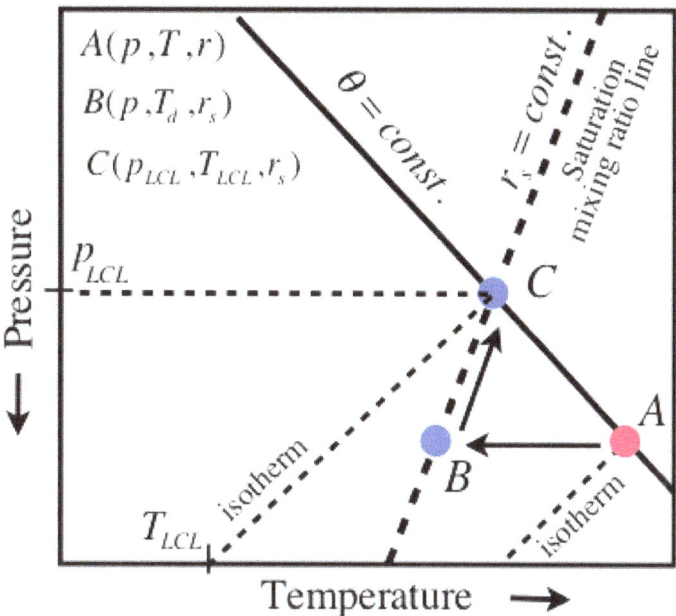

Lifting Condensation Level (LCL): The figure shows the procedure to find the LCL of a parcel undergoing adiabatic ascent along the line θ=constant (isentrope). The isobars are horizontal lines. The parcel is initially at point A having pressure p , temperature T and mixing ratio r . The parcel cools isobarically to become saturated $r_s = r$ its temperature is the Td , the dew point temperature. The point C is the LCL which is the intersection of line r_s = const, and θ = const. Also, note that on this figure, isentropes and sotherms are perpendicular to each other. LCL is also called the isentropic condensation level.

To calculate the lifting condensation level temperature T_l from given values of temperature (T) and the corresponding dew point temperature (T_d), a sufficiently accurate mathematical formula due to Bolton (1980) is

$$\frac{1}{T_l} - \frac{1}{T_d} + 1.266 \times 10^{-3} ln\left(\frac{T_l}{T_d}\right) = 0.514 \times 10^{-3} ln\left(\frac{T}{T_d}\right)$$

In the above formula all temperatures are absolute. However, eq. (4.18a) can only be solved iteratively to obtain T_l. Bolton has also given a simple formula that explicitly gives the ifting condensation level temperature, which reads

$$T_l = \frac{2840}{3.5\,lnT - ln\,e - 4.805} + 55$$

In too all temperatures are in Kelvin.

Wet-bulb temperature: It is measured with a thermometer covered with a moist cloth. The definition of the dew point temperature (T_d) and the wet-bulb temperature (T_w) both involve cooling of a hypothetical air parcel to saturation, but there is a difference. At T_d, air has been cooled to become saturated, i.e. its mixing ratio is r_s. However, the mixing ratio of air as its temperature tends to attain the wet bulb temperature is r; generally $T_w < T$, but equality will apply when air is fully saturated. Usually $T_d < T_w < T$; the thumb rule is that Tw is the arithmetic mean of T and Td ; that is , T_w=0.5 (T + T_d). We shall be using the following terms frequently.

Latent heat of vaporization (L_v) = 2.25 x 10^6 J kg^{-1} (liquid to vapour) at 1 atm and 100°C

Latent Heat of Condensation (L_v) = 2.50 x 10^6 J / kg; (vapour to liquid) at 0°C.

Latent Heat of Melting (L_m) = 3.34 x 10^5 J / kg. (ice to liquid)

Equivalent Potential Temperature

When air is lifted vertically, its ascent will be just like dry air until it reaches saturation; further lifting of the saturated air will result in water vapour conversion to liquid water with the release of latent heat, which must be added in the system. The analysis becomes complex as water may stay in the system or the liquid drops may fall out of the system. In the latter case, the process is pseudo adiabatic, because it is not reversible. Also, heat released adds buoyancy to the rising moist saturated air to further lift it, which may again result in converting vapour to liquid. Consider an amount of heat dQ released due to conversion of water vapour to liquid in the first law of thermodynamics in the form given by eqn. which can be easily put in the entropy from for further analysis. The equation reads as

$$dQ = C_p dT - \frac{dp}{\rho}, which\ implies\ \frac{dQ}{T} = C_p \frac{dT}{T} - R_d \frac{dp}{p}$$

Take logarithm on both sides and then differentiate, $\theta = T \left(\frac{p_0}{p} \right)^{R/C_p}$, then we obtain

$$C_p \frac{d\theta}{\theta} = C_p \frac{dT}{T} - R_d \frac{dp}{p}$$

In eqnuation ofthe term on the right is equal to the entropy of the system therefore

$$C_p \frac{d\theta}{\theta} = \frac{dQ}{T}$$

In view of the potential enthalpy ($C_p \theta$) is taken as a variable instead of θ in the present day global models especially the unified models of atmosphere. It is now required to calculate dQ. When the air parcel is on its 'upward' journey it will first become saturated and any further rise of the now saturated air will condense out dr_s (kg/kg of water), the latent heat released dQ is given by

$$dQ = -L_v dr_s$$

Substituting from equations we obtain,

$$C_p \frac{d\theta}{\theta} = -\frac{L_v}{T} dr_s$$

$$d(\ln\theta) = -\frac{L_v dr_s}{C_p T} \approx -d \left(\frac{L_v r_s}{C_p T} \right)$$

However, it should be proved that $\frac{L_v dr_s}{C_p T} \approx d \left(\frac{L_v r_s}{C_p T} \right)$. Indeed,

$$d\left(\frac{L_v r_s}{C_p T}\right) = \frac{L_v}{C_p}\left[\frac{dr_s}{T} - r_s \frac{dT}{T^2}\right] = \frac{L_v}{C_p T}\left[dr_s - r_s \frac{dT}{T}\right] = \frac{L_v dT}{C_p T}\left[\frac{dr_s}{dT} - \frac{r_s}{T}\right]; \text{ if } \frac{d\,T}{T} << \frac{d\,r_s}{r_s}$$

then $\dfrac{d\,r_s}{d\,T} >> \dfrac{r_s}{T}$

$$d\left(\frac{L_v r_s}{C_p T}\right) \simeq \frac{L_v}{C_p T}\frac{dr_s}{dT}dT = \frac{L_v}{C_p T}dr_s \text{ (neglecting } r_s / T \text{ in the above expression)}$$

Integrating the equation, we obtain the following result

$$ln\theta = -\frac{L_v r_s}{C_p T} + c$$

The constant c is to be evaluated; at low temperatures as $\dfrac{r_s}{T} \to 0, \theta \to \theta_{es} \Rightarrow c = ln\theta_{es}$. Thus,

$$\frac{\theta}{\theta_{es}} \simeq \exp\left\{-\frac{L_v r_s}{C_p T}\right\}; \text{ and } \theta_{es} \simeq \theta \exp\left\{\frac{L_v r_s}{C_p T}\right\}$$

The quantity θ_{es} is called the *saturated equivalent potential temperature*. Accordingly, we can define the equivalent *potential temperature* θ_e as follows,

$$\theta_e = \exp\left\{\frac{L_v r}{C_p T}\right\} = \theta \exp\left\{\frac{L_v q}{C_p T}\right\}; q = specific\ humidity$$

The potential temperature θ is conserved during adiabatic transformations; but equivalent potential temperature θ_e is a conserved quantity during both dry and saturated adiabatic processes. We may define θ_e of a parcel as the potential temperature of the parcel when all its water vapour has condensed out so that its mixing ratio r is zero. To find θ_e the air parcel is first lifted adiabatically so that all the water vapour has condensed and fallen out during its upward journey, and release of latent heat is added to the parcel; then the air parcel is brought back adiabatically downward to *1000 hPa*. The temperature thus attained is the equivalent potential temperature θ_e of the parcel. The equivalent wet-bulb potential temperature can be obtained from wet-bulb temperature using the Poisson's equation. Both θ_e and θ_w (wet-bulb) provide equivalent information for a rising or sinking moist parcel.

Dry Static Energy (DSE): Sum of enthalpy and the potential energy φ;
$$DSE = C_p T + \Phi$$

Moist Static Energy (MSE): Sum of enthalpy, potential energy and latent heat content (Lvq) of a moist parcel. Specific humidity q is close to the mixing ratio r and MSE is defined as,
$$MSE = C_p T + \Phi + L_v q$$

Both DSE and MSE are conserved quantities and play a key role in the vertical motion of a parcel.

When unsaturated air is lifted adiabatically, the enthalpy (C_{p} T) is converted to potential energy (Φ) and the latent heat content remains unchanged during the upward journey; therefore, so long the parcel remains unsaturated, DSE is the key variable in this process. However, when air becomes saturated and further lifted adiabatically, there is an exchange in all the three terms appearing in the expression for MSE. That is, during the upward journey of saturated air, the potential energy (Φ) increases, while the enthalpy (C_pT) and the latent heat content (L_v q) both decrease in such a manner that their sum remains unchanged. Hence, MSE is an important conserved variable for the treatment of clouds in numerical models. Let s represent DSE and h the MSE of a system, then conservation of s and h implies that during adiabatic ascent ds=0 and during saturated adiabatic ascent dh=0,

$$ds = 0 \Rightarrow d(C_p T + \Phi) = 0 \; ; which\ gives\ \frac{dT}{dz} = -\frac{g}{C_p} = -\Gamma_d$$

While for saturated adiabatic ascent of a parcel,

$$dh = 0 \Rightarrow d(C_p T + \Phi + L_v q) = 0$$

Since the specific heat of air change with the addition of moisture, the lapse rate of moist air (Γ_m) will differ from Γ_d in accordance to the water vapour mixing ratio r of the air. For moist air $C_p = C_{pd}(1 + 0.87\ r)$, with C_{pd} (1005 J kg^{-1}K^{-1}) therefore we have

$$\Gamma_m = \frac{\Gamma_d}{1 + 0.87r} \approx \Gamma_d[1 - 0.87r]$$

Entropy of the Dry Air:

The form of the first law of thermodynamics together with the second law of thermodynamics are used to derive an expression for the change in the entropy of the dry air when an amount dQ of heat is added to the system. Taking the first law of thermodynamics,

$$\frac{dQ}{T} = C_p \frac{dT}{T} - R_d \frac{dp}{p}$$

From the second law of thermodynamics, increase in entropy η of the system is defined as

$$d\eta \geq \frac{dQ}{T}$$

But, if heat exchanges reversibly then equality in (4.29) will hold and we have

$$d\eta = \frac{dQ}{T}$$

Thus on combining equation, and using we get

$$d\eta = C_p \frac{dT}{T} - R_d \frac{dp}{p} = C_p \frac{d\theta}{\theta}$$

Thus, eqn. relates the change in entropy dη to change in temperature and pressure of the system. On integrating, we get the entropy of the system as

$$\eta = C_p ln\theta + \eta_1$$

Here η_1 is a constant. Hence, the logarithm of the potential temperature θ gives the entropy of the air, and enables us to *interpret* entropy of the dry air in terms of the well-understood concepts of the potential temperature. The entropy θ or and T constitute the coordinates of a tephigram: θ as the ordinate and T as the abscissa; thus is entropes and isotherms are perpendicular to each other in a tephigram.

Vapour Pressure of Water

The vapor pressure of water is the pressure at which water vapor is in thermodynamic equilibrium with its condensed state. At higher pressures water would condense. The water vapor pressure is the partial pressure of water vapor in any gas mixture in equilibrium with solid or liquid water. As for other substances, water vapor pressure is a function of temperature and can be determined with Clausius–Clapeyron relation.

Vapour pressure of water (0–100 °C)				
T, °C	*T*, °F	*P*, kPa	*P*, torr	*P*, atm
0	32	0.6113	4.5851	0.0060
5	41	0.8726	6.5450	0.0086
10	50	1.2281	9.2115	0.0121
15	59	1.7056	12.7931	0.0168
20	68	2.3388	17.5424	0.0231
25	77	3.1690	23.7695	0.0313
30	86	4.2455	31.8439	0.0419
35	95	5.6267	42.2037	0.0555
40	104	7.3814	55.3651	0.0728
45	113	9.5898	71.9294	0.0946
50	122	12.3440	92.5876	0.1218
55	131	15.7520	118.1497	0.1555
60	140	19.9320	149.5023	0.1967
65	149	25.0220	187.6804	0.2469
70	158	31.1760	233.8392	0.3077
75	167	38.5630	289.2463	0.3806
80	176	47.3730	355.3267	0.4675
85	185	57.8150	433.6482	0.5706
90	194	70.1170	525.9208	0.6920
95	203	84.5290	634.0196	0.8342
100	212	101.3200	759.9625	1.0000

Approximation Formulas

The saturated vapor pressure of water may be approximated by the following relations (in order of increasing accuracy):

$$P \text{ (mmHg)} = \exp\left(20.386 - \frac{5132}{T}\right),$$

where P is the vapor pressure in mmHg and T is the temperature in kelvins.

- Using the Antoine equation

$$\log_{10} P = A - \frac{B}{C + T},$$

where the temperature T is in degrees Celsius and the vapor pressure P is in torr. The constants are given as

A	B	C	T_{min}, °C	T_{max}, °C
8.07131	1730.63	233.426	1	99
8.14019	1810.94	244.485	100	374

- Using the Tetens equation

$$P = 0.61078 \exp\left(\frac{17.27T}{T + 237.3}\right),$$

where temperature T is in degrees Celsius (°C) and vapor pressure P is in kilopascals (kPa)

- Using the Buck equation.

$$P = 0.61121 \exp\left(\left(18.678 - \frac{T}{234.5}\right)\left(\frac{T}{257.14 + T}\right)\right)$$

where T is in °C and P is in kPa.

Accuracy of Different Formulations

Here is a comparison of the accuracies of these different explicit formulations, showing vapour pressures in kPa, calculated at six temperatures with their % error from the table values of Lide (2005):

T (°C)	P (Table)	P (Eq 1)	P (Antoine)	P (Tetens)	P (Buck)
0	0.6113	0.6593 (+7.85%)	0.6056 (-0.93%)	0.6108 (-0.09%)	0.6112 (-0.01%)
20	2.3388	2.3755 (+1.57%)	2.3296 (-0.39%)	2.3399 (+0.05%)	2.3383 (-0.02%)
35	5.6267	5.5696 (-1.01%)	5.6090 (-0.31%)	5.6289 (+0.04%)	5.6268 (+0.00%)
50	12.344	12.065 (-2.26%)	12.306 (-0.31%)	12.354 (+0.08%)	12.349 (+0.04%)
75	38.563	37.738 (-2.14%)	38.463 (-0.26%)	38.718 (+0.40%)	38.595 (+0.08%)
100	101.32	101.31 (-0.01%)	101.34 (+0.02%)	102.43 (+1.10%)	101.31 (-0.01%)

So the simple unattributed formula and the Antoine equation are reasonably accurate at 100 °C, but quite poor for lower temperatures above freezing. Tetens is much more accurate over the range from 0 to 50 °C and very competitive at 75 °C, but Antoine's is superior at 75 °C and above. The unattributed formula must have zero error at around 26 °C, but is of very poor accuracy outside a very narrow range. Tetens' equations are generally much more accurate and arguably simpler for use at everyday temperatures (e.g., in meteorology). As expected, Buck's equation for $T > 0$ °C is significantly more accurate than Tetens, and its superiority increases markedly above 50 °C, though it is more complicated to use. The Buck equation is reportedly even superior to the Goff-Gratch equation.

Numerical Approximations

For serious computation, Lowe (1977) developed two pairs of equations for temperatures above and below freezing, with different levels of accuracy. They are all very accurate (compared to Clausius-Clapeyron and the Goff-Gratch) but use nested polynomials for very efficient computation. However, there are more recent reviews of possibly superior formulations, notably Wexler (1976, 1977), reported by Flatau et al. (1992).

Graphical Pressure Dependency on Temperature

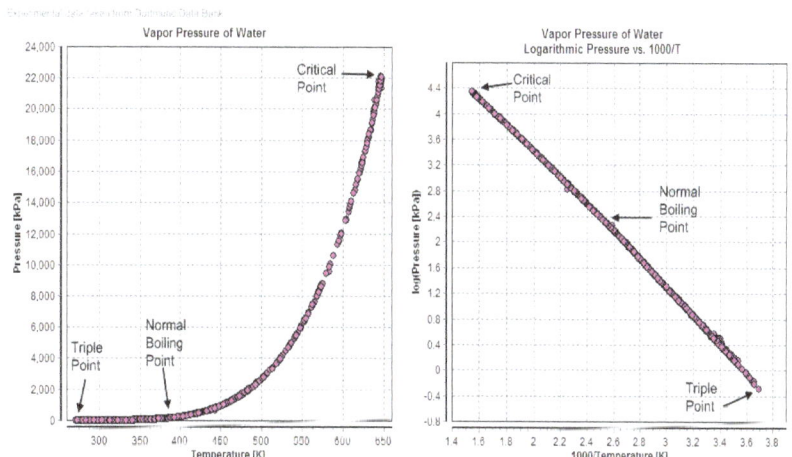

Vapour pressure diagrams of water; data taken from Dortmund Data Bank. Graphics shows triple point, critical point and boiling point of water.

Clausius-Clapeyron Equation

It is one of the most important equations, which is necessary while discussing the water vapour conversion to liquid and its thermodynamic effects in atmosphere. When air is saturated, an equilibrium condition at a given temperature prevails; that is, the number of molecules due to evaporation leaving the liquid phase equals the number of molecules that return to liquid phase due to condensation from the vapour phase. Evaporation takes place when molecules from the surface of water are breaking away as vapour molecules; condensation occurs when vapour phase molecules collide with the liquid water surface and stick to it. Since the kinetic energy of vapour molecules is more than those of liquid, evaporation requires supply of heat to particles at the water surface.

In other words, heat is required to change the liquid phase to vapour. To convert a unit mass of water (phase 1) into vapour (phase 2), the amount of heat required is L_v joules ($L_v = 2.25 \times 10^6$ J kg^{-1}) which is,

$$L_v = \int_{q1}^{q2} dQ = \int_{u1}^{u2} du + \int_{a1}^{a2} p d\alpha \qquad \text{..(1)}$$

For saturated air, p=e_s and it is constant throughout the process; therefore, we have from (1)

$$L_v = u_2 - u_1 + e_s(\alpha_2 - \alpha_1) \qquad \text{...(2)}$$

Since temperature T is also constant during the phase change process, we may also write

$$L_v = \int_{q1}^{q2} dQ = T \int_{q1}^{q2} \frac{dQ}{T} + T \int_{\eta1}^{\eta2} d\eta \Rightarrow L_v = T(\eta_2 - \eta_1) \qquad \text{.........................(3)}$$

Equating the expression in (2) with that in (3), we obtain

$$u_2 - u_1 + e_s(\alpha_2 - \alpha_1) = T(\eta_2 - \eta_1) \qquad \text{.....................................(4)}$$

The above expression on rearrangement gives

$$u_1 + e_s\alpha_1 - T\eta_1 = u_2 + e_s\alpha_2 - T\eta_2 \qquad \text{..(5)}$$

Let us introduce the Gibbs Free Energy function (or simply the Gibbs Function) G as

$$G = u + e_s\alpha - T\eta \qquad \text{...(6)}$$

With the definition of the Gibbs function, eqn. (5) can be written as

$$G_1 = G_2 \qquad \text{...(7)}$$

We thus infer from (5) and (7) that in an isothermal (T=constant), isobaric (p=constant) phase change, the Gibbs function G, remains constant. The Gibbs function G depends on temperature and pressure, but it *remains constant* during phase transition.

In order to determine the dependence of G on pressure and temperature, we differentiate eqn. (6) to obtain,

$$dG = du + e_s d\alpha + \alpha de_s - Td\eta - \eta dT \qquad \text{..(8)}$$

Now $dQ = du + e_s d\alpha$ and $dQ = Td\eta$ and $du + e_s d\alpha - Td\eta = 0$ thus in (4.38) and it becomes,

$$dG = \alpha de_s - \eta dT \qquad \text{..(9)}$$

Because G is same for both phases $dG_1 = dG_2$ gives

$$\alpha_1 de_s - \eta_1 dT = \alpha_2 de_s - \eta_2 dT \qquad \text{...(10)}$$

$$\frac{de_s}{dT} = \frac{\eta_2 - \eta_1}{\alpha_2 - \alpha_1} \quad \text{..(11)}$$

Substituting for η2 -η1 from eq. (13) in (11) we obtain the following relation

$$\frac{de_s}{dT} = \frac{L_v}{T(\alpha_2 - \alpha_1)} \quad \text{..(12)}$$

The equation (12) is the *Clausius-Clapeyron equation,* which expresses the change in saturation pressure e_s that results from a dT change in temperature. Under ordinary atmospheric conditions, vapour phase specific volume is much larger than that of the liquid phase, i.e. $\alpha_2 \gg \alpha_1$, and moreover, water vapour also behaves like an ideal gas; hence, we can neglect the term α_1 in the equation (12) to obtain the final form of the Clausius-Clapeyron equation after substituting for as $\alpha_2 = R_v T / e_s$

$$\frac{de_s}{dT} = \frac{L_v}{T\alpha_2} \Rightarrow \frac{de_s}{dT} = \frac{L_v e_s}{R_v T^2} \quad \text{.......................................(13)}$$

At temperatures below 0°C, eq. (13) expresses the saturation vapour pressure of super-cooled liquid water. However, (13) needs modification for expressing saturation vapour pressure of ice. For this case, latent heat of sublimation L_s is substituted in place of L_v and the change of saturation vapour pressure over ice with temperature is calculated from the following relation,

$$\left(\frac{de_s}{dT}\right)_{ice} = \frac{L_s e_s}{R_v T^2} \quad \text{..(14)}$$

At temperature above 0°C, only liquid water and water vapour are in equilibrium. Hence Clausius-Clapeyron equation is the key equation in phase change.

In eqn. (4.43), R_v=461.55 J kg^{-1}K^{-1} is constant and to a first approximation, if L_v is also taken as a constant then the straightforward integration of (4.43) yields,

$$\ln\left(\frac{e_s}{e_{s0}}\right) = \frac{L_v}{R_v}\left[-\frac{1}{T}\right]_T^T \Rightarrow \ln\left(\frac{e_s}{e_{s0}}\right) = \frac{L_v}{R_v}\left[\frac{1}{T_0} - \frac{1}{T}\right] \quad \text{..........................(15)}$$

In (15), $e_{s0} = e_s(T_0)$ is the value of saturation vapour pressure at the temperature T_0. It is found that e_{s0} = 6.112 hPa for the triple point temperature T_0=273.16 K. We therefore write (15) as

$$e_s = e_{s0}\,\exp\left\{\frac{L_v}{R_v}\left[\frac{1}{T_0} - \frac{1}{T}\right]\right\}; e_{s0} = 6.112 hPa \quad \text{.........................(16)}$$

$$e_s = e_{s0}\,\exp\left(c - \frac{B}{T}\right); C = 19.8313 \quad \text{..(17a)}$$

$$e_s = Ae^{-B/T} \; ; \; A = 2.53345 \times 10^9 \, hpa; \; B = 5417.1181 \, K \qquad \qquad \text{............(17b)}$$

At subfreezing temperatures, saturation vapour pressures for vapour and ice can be compared from the following expression

$$\frac{e_s}{e_{si}} = \exp\left\{\frac{L_f}{RT_0}\left(\frac{T_0}{T} - 1\right)\right\}; \; L_f = L_s - L_v \qquad \qquad \text{............(18a)}$$

Numerically, a good approximation to this equation is the following relation,

$$\frac{e_s}{e_{si}} = \left(\frac{273}{T}\right)^{266} (T \text{ is in Kelvin}) \qquad \qquad \text{............(18b)}$$

One can compute $e_s(T)$ starting with $e_{so}(T_o)$ and $T_o = 273.16$ K using the Clausius-Clapeyron relation. One can create a look-up table for every $\delta T = 1$ K increment in temperature in the above formulae to speed up the calculations. More accurate computation of $e_s(T)$ and $e_{si}(T)$ in hPa units have been computed by Tetens' formulae (with T_0 and T_o in Kelvin) as given below,

$$e_s = e_{so} \exp\left\{17.502 \frac{T - T_0}{T - 32.19}\right\} (hPa); \; \textit{over liquid water} \qquad \qquad \text{............(19a)}$$

$$e_{si} = e_{sio} \exp\left\{22.587 \frac{T - T_0}{T + 0.7}\right\} (hPa); \; \textit{over ice} \qquad \qquad \text{............(19b)}$$

A more accurate formula for saturated vapour pressure for $e_s(T)$ was derived by Bolton (The computation of equivalent potential temperature. Monthly Wea. Rev. vol.108, 1980). The empirical formula with 0.1% accuracy over the temperature range -30°C $\leq$ T $\leq$ 35°C, is as follows

$$e_s(T) = 6.112 \exp\left\{\frac{17.67\,T}{T + 243.5}\right\} (e_s \text{ in } hPa; T \text{ in } ^0C) \qquad \qquad \text{............(19c)}$$

References

- Minder, JR; Mote, PW; Lundquist, JD (2010). "Surface temperature lapse rates over complex terrain: Lessons from the Cascade Mountains". J. Geophys. Res. 115: D14122. Bibcode:2010JGRD..11514122M. doi:10.1029/2009JD013493

- Mark Zachary Jacobson (2005). Fundamentals of Atmospheric Modeling (2nd ed.). Cambridge University Press. ISBN 0-521-83970-X

- Stewart, Robert H. (September 2008). "6.5: Density, Potential Temperature, and Neutral Density". Introduction To Physical Oceanography (pdf). Academia. pp. 83–88. Retrieved March 8, 2017

- Todd S. Glickman (June 2000). Glossary of Meteorology (2nd ed.). American Meteorological Society, Boston. ISBN 1-878220-34-9

- Doswell, Charles A.; Rasmussen, Erik N. (1994). "The Effect of Neglecting the Virtual Temperature Correction on CAPE Calculations". Weather and Forecasting. 9 (4): 625–629. doi:10.1175/1520-0434(1994)009<0625:-TEONTV>2.0.CO;2

- Dr. James T. Moore (Saint Louis University Dept. of Earth & Atmospheric Sciences) (August 5, 1999). "Isentropic Analysis Techniques: Basic Concepts" (pdf). COMET COMAP. Retrieved March 8, 2017

- Stull, Roland B. (2001). An Introduction to Boundary Layer Meteorology (1st ed.). Kluwer Academic Publishers. ISBN 90-277-2769-4

- Camargo, Suzana J.; Sobel, Adam H.; Barnston, Anthony G.; Emanuel, Kerry A. (2007). "Tropical cyclone genesis potential index in climate models". Tellus A. 59 (4): 428–443. doi:10.1111/j.1600-0870.2007.00238.x

Ocean Thermodynamics: A Comprehensive Study

Climate depends on the interaction of atmospheric processes with other Earth systems. Temperature and climatic conditions of the landmass is directly affected by the ocean. Atmospheric and oceanic physics is best understood in confluence with the major topics listed in the following chapter.

Ocean Thermodynamics

Oceanography as a systematic science began in 19th century along with meteorology. *"Ocean waters are saline"* refers to the two most remarkable constituents of oceans, viz., *the water and salt*s. The other well known fact *about ocean that it is opaque to all electromagnetic radiation at wavelengths that are used in remote sensing of the atmosphere*. Due to this reason, *in situ* observational devices are used in expeditions. However, in the early stage of oceanography, progress was achieved by brave seafarers who waded difficult waters and confronted cyclones and other impediments of voyage; if frustrated renewed their dream to reach remote destinations, and finally succeeded in using their inventions to collect precious observations which the mankind treasures now in the form of comprehensive knowledge of intricate oceanic phenomena. It took them therefore several years of hard labour often in solitude to constitute a global picture of ocean circulation, horizontal and vertical thermal structure, and distribution of various scalar quantities such as salinity, oxygen content etc. from these time staggered ship observations. As a result, the ocean was regarded in a state of steady, large-scale flow devoid of turbulent motions. Today, this picture has completely changed as the oceanic circulations are dominated by turbulent eddies of different temporal and spatial scales.

Above all, ocean is the main reservoir of water in the atmosphere, holds the key to long range forecasting of weather and to mitigate the much feared climate change due to continual increase of anthropogenic activity which has exhausted most of the land resources. *Oceans are the future resource bed of human needs*, and an understanding of ocean physics is therefore very essential in exploiting the treasures within the ocean and those at its bottom.

Three Responsible Factors and Three Eras of Oceanographic Research

The developments in oceanography have been slow but *three main factors* led to the present state of knowledge of oceans; these are: *(i) sea explorations that arose from human curiosity, (ii) urgent need of depth measurements for engineering purposes, and (iii) the life in the depths of sea.* The corresponding *three eras* of oceanographic research are: *(i) three–dimensional exploration of the physical, chemical, biological and geological environment of the seas may be regarded as the first era; (ii) the second era spans between the World War I and World War II; and (iii) the use of mod-*

ern oceanographical techniques and instruments marks the beginning of the third era in the exploration of oceanic waters. The present day instruments such as neutrally buoyant floats (buoys), network of Argo floats and a number of sensors and cameras deployed on satellites constantly survey the structure and temporal behaviour of drifting water masses on the globe. These observations from a variety of platforms form the very basis of finding the interrelationship of oceanic changes forced by atmospheric circulation. Drifting buoys and Argo floats are the marvels of technology. Argo floats have been deployed in the open sea to constantly measure ocean currents, temperature, salinity, oxygen contents etc. in deep layers extending from surface to 2000 m at their positions. The drifting buoys map surface currents and produce important information on the sea state. Data from all such platforms are then transmitted to a satellite to reach the users. Argo floats have a mechanism to change their buoyancy and regularly record ocean parameters up to a depth of 2000 m, and they are deployed in all the ocean basins to create a global network of Argo observations.

From the earliest oceanographic explorations – the Indian Ocean Expeditions and the International Cooperative Investigations of the Tropical Atlantic (ICITA) – several nations are cooperating to understand the climate and its change. As a result of this cooperation, the decade long Tropical Ocean Global Atmosphere / Coupled Ocean–Atmosphere Response Experiment (TOGA/CORE), the Indian Ocean Experiment (INDOEX) and several other experiments in the Atlantic and Pacific have been successfully carried out. *"All nations agreeing to unite and cooperate in carrying out one system of philosophical research"* is one such peculiar spectacle of the scientific world, that is becoming larger than ever to make the famous *International Programme on Climate Change (IPCC)* a successful story, where not only the savants and individual experts from different branches of science and engineering are cooperating but also different governments are funding and encouraging research to address the highly sensitive issue of *"global warming"* arising due to anthropogenic activity and excessive industrial emissions. *"Save the Planet"* is a familiar and most loud cry from the environmentalists, but it cannot stop the pace of industrial production that has assured the kind of life humans are enjoying in today's world. This realisation is the main driver of inventions and development of efficient technologies.

The CO_2 is constantly increasing because several new nations with large areas and populations are joining the race to produce more and more by increasing their industrial activity. The trends appear irreversible and solutions to the problem could therefore come from deeper understanding of the processes and if it leads to development of newer technologies to control the runaway increase of CO_2 on this planet, that would surely mean no small fortune to mankind. More so, oceans constitute a major sink of carbon dioxide and cover almost 75% of the earth's surface. Oceans also store vast amount of heat and therefore mollify the extreme climates – cold or hot.

The salinity variations and the pole-to-equator temperature gradients set oceans into thermohaline circulation. Similarly, winds also impart momentum to produce regions of warm waters to move to different geographical locations to produce the notable Gulf Stream and other currents in North Atlantic and Pacific, and various currents in the Indian Ocean already discussed earlier. The changes in the sea surface temperature in turn affect the winds and turbulent exchanges. Similarly winds also play an important role in triggering El Niño in the Pacific Ocean, which is intimately linked to Southern Oscillation in the atmospheric circulation and therefore commonly known as the El Niño Southern Oscillation (ENSO) phenomenon. Monsoon over India is another major circulation where ocean temperatures play a dominant role in its performance and seasonal/annual variability. In essence, a profound understanding of the air-sea interaction holds the key to seasonal weather forecasting, climate predictions and even to diagnosing reasons for climate change. The key point of above discussion is that oceans and atmosphere form a single system. In this context it is apt to quote here W.S. von Arx from his book *An Introduction to Physical Oceanography:* "Because the ocean and atmosphere are so closely interconnected in so many ways and because both are ultimately dependent on solar heating for the energies of their motion, as well as their characteristic properties, it is misleading to discuss either one without the other."

Thermodynamics of Seawater

The key characteristics of ocean waters have been stated in the preceding paragraph. The organized motions in the ocean allow vast amounts of heat storage into different oceanic layers in the vertical. Next, the physical principles controlling the oceanic processes are to be formulated, which require an understanding of the thermodynamics of seawater.

Equation of State : The density ρ of seawater depends on pressure p, temperature T, and salinity S (*g/g or kg/kg*) though for a one-component fluid such as pure water, it a function only of pressure and temperature. The density of seawater as a multicomponent mixture is given by

$$\rho = \rho(p, T, S)$$ ....(1)

For seawater, the equation of state is most appropriately expressed in the differential form

$$\frac{1}{\rho} d\rho = \gamma_T dp - \alpha_T dT + \beta ds$$

....(2)

$$\alpha_T = \frac{1}{\rho} \frac{\partial \rho}{\partial T} \qquad \textit{Coefficient of thermal } \exp \textit{ansion,}$$

$$\gamma_T = \frac{1}{\rho} \left(\frac{\partial \rho}{\partial p} \right) \textit{Isothermal compressibility coefficient}$$

$$\beta = \frac{1}{\rho} \frac{\partial \rho}{\partial S} \qquad \textit{Coefficient of saline contraction}$$

$$\alpha = \frac{1}{\rho} \qquad \textit{Specific volume}$$

An equation of state of sufficient accuracy is essentially needed for the computation of density of seawater, which is the key variable in determining ocean currents by the so–called dynamic methods. An internationally agreed upon equation of state fits the available density measurements with an accuracy of the order of 3.5 x 10^{-6} over the frequently encountered ranges of oceanic pressure, temperature and salinity. This equation reads

$$\rho = \rho(p, T, S) = \frac{1}{\alpha(p, T, S)} = \frac{\rho(o, T, S)}{1 + \left\langle \frac{1}{\alpha} \frac{\partial \alpha}{\partial p} \right\rangle p}$$ ....(3)

The mean compressibility, $\left\langle -\frac{1}{\alpha} \frac{\partial \alpha}{\partial p} \right\rangle$ is defined as

$$-\frac{\alpha(p,T,S)-\alpha(o,T,S)}{\alpha(0,T,S)p}=\frac{1}{K_T(p,T,S)}. \quad\text{.....................................(4)}$$

In (4) $K_T(p,T,S)$ is the mean bulk modulus. One may write using (4) the specific volume α in the form

$$\alpha(p,T,S)=\alpha(o,T,S)\left[1-\frac{p}{K_T(p,T,S)}\right]\text{...............................(5)}$$

The density of surface pressure (p = o) is expressed as

$$\rho(0,T,S)=\frac{1}{\alpha(0,T,S)}=A+B\times S+C\times S^{3/2}+D\times S^2 \quad\text{..........................(6)}$$

Note that pressure at the sea surface p = o is obtained by subtracting atmospheric surface pressure p_s at any point. The bulk modulus $K_T(p,T,S)$ is given by

$$K_T(p,T,S)=E+F\times S+G\times S^{3/2}+(H+I\times S+J\times S^{3/2})p+(M+N\times S)p^2 \quad\text{.....(7)}$$

The temperature T is specified in degrees Celsius (°C); the pressure p in bars, or 10^5Pa; salinity S in "practical salinity units (psu)" which replaces the former units °/oo (parts per thousand). Indeed, several ocean scientists have strengthened the concept of "absolute salinity" in order to improve the accuracy of climate models. The salinity expressed in psu is used to derive the absolute salinity by computing a correction term that is computed from look-up tables (TOES 2010), which is then added to the salinity in conventional units. The equations (3), (5), (6), and (7) thus give the most accurate form of the equation of state of seawater which will give density ρ (kg/m³) and specific volume α (m³/kg) from observed values of salinity (psu), temperature (°C) and pressure (bar) within a standard error of approximately 0.009 kg/m3 over the entire range of oceanic pressures. The seawater has 96.5 % of water content and 3.5% consists of dissolved materials in the form of molecules or ions. Though relatively small but important 3.5% (which is equal to 35 psu) of materials in seawater have a profound impact on ocean currents, formation of deep waters and the ocean stratification. Therefore highly accurate computations of seawater density from the equation of state are necessary. Also note that accuracy of the density of seawater inevitably depends on the accuracy of salinity; for this reason, the relationship between the salinity and the electrical conductivity of seawater has to be highly accurate as the former quantity is determined by the latter. The coefficients A, B, C, ..., N that appear in the equations (1)-(7) are given by Fofonoff (1985) which have been tabulated below. All those coefficients A, B, ..., N are polynomials up to fifth degree. As for example, A is written as

$$A=a_0T^0+a_1T^1+a_2T^2+a_3T^3+a_4T^4+a_5T^5=\sum_{k=1}^{5}a_kT^k \quad\text{...................(8)}$$

The constants a_0, a_1, ..., a_5 that respectively multiply T^k (i.e. the power of T is determined by the index k), are arranged in the column A of the Table and its value can be computed using equation (8). Similarly, the coefficients of various powers of T appearing in other terms, viz., B, C, D, ... , N have been arranged in the correspondingly named column of Table. The blanks should be taken as zero in the table. Thus the polynomial expression of coefficient B will be of order 4; the coefficient C will be expressed by a quadratic; while the coefficient D would be just equal to a constant value

(= 4.8314 x 10^{-4}). The table could be directly used in developing a computer program which would readily yield the density of seawater upon supplying the in situ values of temperature and salinity. The loci of constant density can then be easily plotted for different values of temperature (0-30°C) and salinity (0-40 psu) for understanding its variations over any region.

The density of seawater varies slowly in space and time. A typical value of density of seawater at the surface is $\rho = 1.026 \times 10^3$. The isothermal compressibility coefficient γ_T at a depth 2000 m and temperature 4°C is found a to be of order 10^{-5}; that is,

$$\gamma_T = \frac{1}{\rho}\left(\frac{\partial \rho}{\partial p}\right)_T = 4.3 \times 10^{-10} Pa^{-1} = 43 \times 10^{-6} bar^{-1} \qquad ....(9)$$

Assuming hydrostatic pressure, it is possible to calculate the average fractional increase in density from the term, $\gamma_T \dfrac{dp}{dz}$; thus

$$\gamma_T \frac{dp}{dz} = \frac{1}{\rho}\left(\frac{\partial \rho}{\partial p}\right)_T \left(\frac{dp}{dz}\right) = \gamma_T \rho g = 4.3 \times 10^{-10} \times 1.026 \times 10^3 \times 9.81 = 43 \times 10^{-6} m^{-1}$$

That is, increase in the density of seawater, due to the weight of the overlying mass of fluid, is approximately 4.3 per millionth of meter. The reciprocal of $\gamma_T \dfrac{dp}{dz}$; is called the "scale height" H, i.e.

$$H = \frac{1}{\gamma_T\left(\dfrac{dp}{dz}\right)} = \frac{1}{43 \times 10^{-6}} m \approx 230\, km \qquad ....(10)$$

Hence, the e-fold increase in density due to isothermal density alone shall happen at a depth of 230 km of ocean. By good fortune, ocean depth is only about 8 km, hence considerable mathematical simplifications are possible in expressing density of seawater.

Table: Coefficients in the computations of terms A, B, C etc. in the equation of state for seawater (Fofonoff 1985)

	A	B	C
T^0	+999.842594	+8.24493 e-1	- 5.72466 e-3
T^1	+6.793952 e-2	- 4.0899 e-3	+1.0227 e-4
T^2	- 9.095290 e-3	+7.6438 e-5	- 1.6546 e-6
T^3	+1.001685 e-4	- 8.2467 e-7	
T^4	- 1.120083 e-6	+5.3875 e-9	
T^5	+6.536332 e-9		

	D	E	F
T^0	+4.8314 e-4	+19652.21	+54.6746
T^1		+148.4206	- 0.603459
T^2		- 2.327105	+1.09987 e-2
T^3		+1.360477 e-2	- 6.1670 e-5
T^4		- 5.155288 e-5	

	G	H	I
T⁰	+7.944 e-2	+3.239908	+2.2838 e-3
T¹	+1.6483 e-2	+1.43713 e-3	- 1.0981 e-5
T²	- 5.3009 e-4	+1.16092 e-4	- 1.607 e-6
T³		- 5.77905 e-7	

	J	M	N
T⁰	+1.91075 e-4	+8.50935 e-5	- 9.9348 e-7
T¹		-6.12293 e-6	+2.0816 e-8
T²		+5.2787 e-8	+9.1697 e-10

Such a large value of H means that *average density of the ocean varies slightly even in the deepest portion of the sea*. Further (10) suggests that pressure variations in the equation of state of seawater could be neglected leading to considerable simplifications. Notwithstanding this inference, *even small variations in seawater density are indeed non-negligible. Indeed, if ocean were truly incompressible, the ocean surface would perhaps be 30 m higher than from the actual level*. For this reason, compressibility must be taken into account in temperature and salinity measurements in deep layers of ocean. In many calculations, assuming density constant, except in terms where it appears with acceleration due to gravity (the Boussinesq approximation), leads to considerable simplifications of the governing equations. Ocean is also a stratified fluid medium like atmosphere but there are two distinct differences:

1. In atmosphere, moisture is a key thermodynamic variable which undergoes phase change releasing latent heat which drives the atmospheric circulation. However, salinity is an important variable of ocean thermodynamics, but for sure, it cannot serve as a counterpart of moisture in the atmosphere as it does not undergo phase change. It certainly depresses the freezing point of seawater and increases density.

2. All oceans are laterally bounded except in the Southern Ocean that extends around the globe and waters can pass through the Drake Passage, a 600 km narrow gap between the tip of South America and Antarctic Peninsula.

Sigma-t and potential density: An immediate simplification in (2) is practically straightforward when pressure changes are assumed of no consequence. Hence by putting dp=0 in the differential form of the equation of state of seawater (2), we get

$$\frac{1}{\rho} d\rho = -\alpha_T dT + \beta dS \qquad \dots\dots\dots\dots\dots\dots\dots\dots\dots\dots\dots\dots\dots\dots(11)$$

If the deviations are called from a standard reference state then (6.11) reduces to an algebraic equation

$$\rho = \rho_0 + \rho_{ref}[-\alpha_T(T - T_o) + \beta(S - S_0)] \qquad \dots\dots\dots\dots\dots\dots\dots(12)$$

That is, if one may assume density independent of pressure, a highly simplified equation of state may therefore be written as

$$\sigma_t = \rho' = \rho'_0 + \rho_{ref}\left[-\alpha_T(T-T_o)+\beta(S-S_0)\right] \quad\text{...(13)}$$

$$\alpha_T = \frac{1}{\rho}\frac{\partial\rho}{\partial T}, \quad \beta = \frac{1}{\rho}\frac{\partial\rho}{\partial S}, \quad \rho' = \rho - 1000(kg\,m^{-3}), \quad \rho'_0 = \rho'_0(T_0, S_0)$$
$$\text{....................(14)}$$

The *density anomaly* σ_t is the difference between actual density and the reference density r ref = 1000 kg m⁻³. Since the variations in the density of seawater are less than 7%, the variable σ_t is a useful quantity in the ocean analysis. Further the effect of pressure on density is eliminated with the use of reference density ρ_{ref} in the definition of σ_t which is just a function of temperature (*T*) and salinity (*S*). On the T-S diagram, equal sigma-t (σ_t) lines are curved. One may notice in Figure that the slope of the potential density curves increases with decreasing temperatures. Further, with the help of T-S diagrams, density profiles from observing stations in the ocean can be easily analysed to infer about the type of waters and their geographical location.

From the equation of state (2) in the differential form, it is obvious that variations in density of ocean waters arise mainly from evaporation and precipitation, as they are associated with important changes in both temperature and salinity. Another key mechanism for such variations is the mixing of water masses at different temperatures and salinity. In essence, heavier cold saline waters bring into focus the marginal seas where most of the deep waters are formed. The density anomalies mainly arise in the shallow coastal waters due to mixing of seawater with freshwater contributed by the runoff from big river systems and melting of ice.

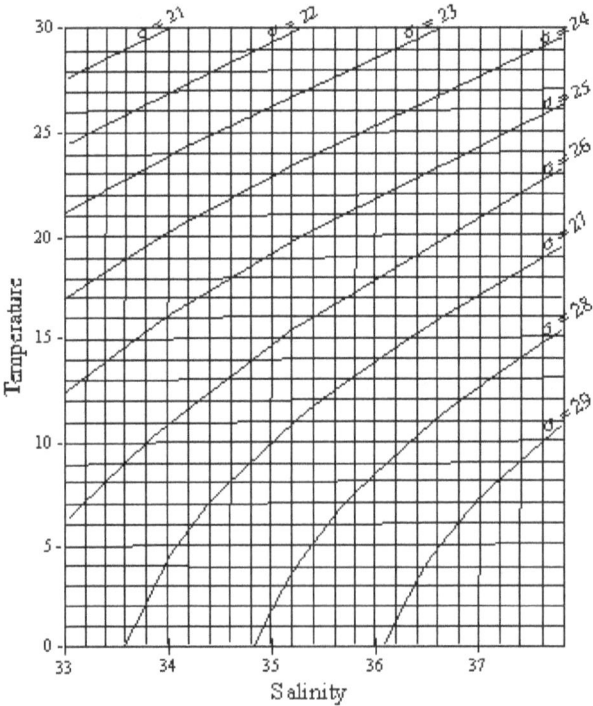

Temperature-salinity (T-S) diagram: The equal density anomaly curves
are labelled from σ_1= 21,........,29 kg m⁻³ .
The subscript "t" of σ_t has been dropped while labelling the curves for clarity of symbols. The temperature scale ranges from 0-30 °C with an increment of 1°C.Salinity ranges from 33-38 psu with an increment of 0.2 psu.

In a stratified ocean, the hydrostatic equation is applicable, hence

$$\frac{\partial \bar{p}}{\partial z} = -g\bar{\rho}(z)$$

....(15)

Concept of a parcel: The major ocean basins are Atlantic, Indian Ocean and Pacific. The waters of these basins in different layers are characterized by salinity and temperature. The oceans are dynamic and water masses constantly move from one basin to another along the ocean currents forced by winds and gradients of temperature and salinity (thermohaline circulation). The temperature and salinity profiles in each basin determine its thermohaline structure and force meridional thermohaline circulation in which dense abyssal waters are formed in high latitudes. After traversing a tortuous route (depicted by the Conveyer Belt) in the depths of different basins, ocean waters again reach to the surface. Ocean mixing due to roughness of bottom topography plays an important role in changing the characteristics of deep waters during their movement due to thermohaline circulation. The concept a parcel is therefore important in assessing the conversion of dense abyssal waters to light water masses that traverses back to surface. A parcel of water mass of a typical basin preserves its characteristics and remains insulated during its long journey in the ocean. The ocean structure is thus directly inferred from the vertical movements of the parcels of water masses. The static stability of the ocean can be related to the vertical gradients of density which depends upon salinity and temperature.

Stability of ocean stratification: The static stability Γ_s is defined as

$$\Gamma_s = -\frac{1}{\rho}\frac{\partial \bar{\rho}}{\partial Z}$$

....(16)

From the equation of state of seawater (2) it is possible to write $\dfrac{1}{\rho}\dfrac{\partial \bar{\rho}}{\partial Z}$ as

$$\frac{1}{\rho}\frac{\partial \bar{\rho}}{\partial Z} = -\gamma_T \bar{\rho} g - \alpha_T \frac{\partial \bar{T}}{\partial Z} + \beta \frac{\partial \bar{S}}{\partial Z}$$

....(17)

Thus, the background temperature $\bar{T}$, salinity $\bar{S}$ and isothermal compressibility coefficient (γ_T) are required in determining the vertical density profile in the ocean. While in the upper layers (especially in the mixed layer) and seasonal thermocline, temperature and salinity effects are most pronounced and highly variable; but in deeper regions, ocean is neutrally stratified or very slightly stable. Thus, in deeper regions, density and temperature changes are mostly due to adiabatic compression of seawater by the pressure of the overlying fluid.

In order to derive the stability condition for vertical velocities, consider an isolated small parcel of fluid which may expand freely if pressure of the surrounding fluid decreases. Let $\rho_0(z)$ denote the density at the centre of the parcel at an undisturbed position z; T_0 and S_0 are respectively the temperature and salinity of the parcel at this position. If a parcel were raised to a height z+δz slowly without disturbing the horizontal stratification of the surrounding fluid, the hydrostatic pressure acting on the parcel would change to δp (= - ρ_0gδz) for small δz and the adiabatic expansion would change the density of the parcel. Thus,

$$\rho_0(Z+\delta Z)=\rho_0(Z)+\frac{\partial \rho_0}{\partial Z}\delta Z=\rho_0(Z)\left(\frac{\partial \rho}{\partial p}\right)\left(\frac{\partial p}{\partial Z}\right)\delta Z \quad\quad\quad\quad\quad \text{............................(18)}$$

$$\rho_0(Z+\delta Z)=\rho_0(Z)+\left(\frac{1}{\rho_0}\frac{\partial \rho}{\partial p}\right)\left(\frac{\partial p}{\partial Z}\right)\rho_0\delta Z=\rho_0(Z)-\gamma_a\rho_0^2\delta Z$$

$\gamma_a=\frac{1}{\rho_0}\frac{\partial \rho}{\partial p}$ is the adiabatic compressibility coefficient. From thermodynamics, it is well known that the volume of a fluid will decrease by greater magnitude in case of isothermal compression than adiabatic compression, therefore $\gamma_0 < \gamma_T$. The dynamics of displaced parcels, as a consequence, depends mainly upon the difference $\gamma_T - \gamma_a$ or on the quantity c_1 which has same dimensions and magnitude of speed of sound in the liquid:

$$c_1=\frac{1}{\sqrt{\rho_0(\gamma_T-\gamma_a)}} \Rightarrow \rho_0(\gamma_T-\gamma_a)=\frac{1}{c_1^2} \quad\quad\quad\quad \text{.............................(19)}$$

Buoyancy of seawater parcels: The upward directed pressure gradient force acting on a parcel at z+δz, according to Archimedes (212 BC) , is equal to the local volume of the parcel multiplied by the local density ρ'(z+δz) of the surrounding fluid. The downward force of gravity on the parcel is equal to the product of volume and the density ρ_0(z+δz) of the parcel. The local volume of a parcel of unit mass is given by α_0= 1/ρ_0 . Thus, the buoyancy of a parcel depends upon the upward directed (pressure gradient) force acting on the volume α_0 and the weight of the parcel which are as follows.

$$\textit{Upward directed force on volume } \alpha_0 = g\rho(z+\delta z).\alpha_0$$
$$\textit{Downward acting force due to gravity} = g\rho_0(z+\delta z).\alpha_0$$

Hence the buoyancy force acting on the parcel of volume α_0 is given by the difference of the upward and downward forces acting on the parcel,

$$(\Delta F)_{buoy}=g[\overline{\rho}(Z+\delta Z)-\rho_0(Z+\delta Z)]\alpha_0 \quad\quad\quad\quad \text{.............................(20)}$$

The expression (20) also can also be interpreted as the acceleration of a unit mass in the upward direction, which can be written as

$$\frac{g}{\rho_0}[\overline{\rho}(Z+\delta Z)-\rho_0(Z+\delta Z)]=\frac{g}{\rho_0}\left|\frac{\partial \overline{\rho}}{\partial Z}\right|\delta Z+\overline{\rho}(Z)-\rho_0(Z)+\gamma_a\rho_0^2\delta Z \quad\quad \text{..................(21)}$$

$$\text{Or } \frac{g}{\rho_0}[\overline{\rho}(Z+\delta Z)-\rho_0(Z+\delta z)]=g\left[\frac{1}{\rho_0}\frac{\partial \overline{\rho}}{\partial Z}+\gamma_a\rho_0 g\right]\delta Z$$

Initially, the parcel has started from level z, where the density of the stratification and of the parcel are same, i.e. $\overline{\rho}(Z)=\rho_0(Z)$; this leads us to the final form (21). The acceleration per unit mass of a parcel moving upward by a displacement (δz) is given by $\frac{d^2(\delta Z)}{dt^2}$, which can be equated to the right hand side of (21) to yield the following equation,

$$\frac{d^2(\delta Z)}{dt^2} = g\left[\frac{1}{\rho_0}\frac{\partial \bar{\rho}}{\partial Z} + \gamma_a \rho_0 g\right]\delta Z = -N^2 \delta Z$$

....(22)

$$N^2 = -g\left[\frac{1}{\rho_0}\frac{\partial \bar{\rho}}{\partial Z} + \gamma_a \rho_0 g\right] = g\left[\gamma_T \bar{\rho}g + \alpha_T\frac{\partial \bar{T}}{\partial Z} - \beta\frac{\partial S}{\partial Z} - \gamma_a \rho_0 g\right]$$

............................(23)

Taking $\bar{\rho}(Z) = \rho_0(Z)$ and using eqn. (19), the above expression for N² becomes,

$$N^2 = -g\left[\alpha_T\frac{\partial \bar{\rho}}{\partial Z} - \beta\frac{\partial \bar{S}}{\partial Z} + \frac{g}{c_1^2}\right]$$

....(24)

Thus equation (22) implies that parcel oscillates with Brunt-Väisälä frequency N. An important deduction from equation (24) is now possible. One may define the isohaline fluid that has the stratification with $\frac{\partial \bar{S}}{\partial Z} = 0$. In such a fluid the "restoring force" may vanish, i.e. N²= 0, which in turn implies that

$$\alpha_T\frac{\partial \bar{T}}{\partial Z} + \frac{g}{c_1^2} = 0; \ \ and \ gives \ \frac{\partial \bar{T}}{\partial Z} = -\frac{g}{\alpha_T c_1^2}$$

....(25)

Hence, in an isohaline fluid, the restoring force will vanish (i.e. the parcel will not oscillate) if the basic temperature gradient is equal to -g/($\alpha_T c_1^2$). The equation (25) implies that in an isohaline fluid stratification, the temperatureT_0(Z+dZ) of the displaced parcel must be equal to the temperature $\bar{T}(Z + \delta Z)$ of the surrounding fluid. This is because the densities as well as pressures are equal when N=0. Therefore, the adiabatic temperature change δT_0 of a parcel is,

$$\delta T_0 = \left(\frac{\partial \bar{T}}{\partial Z}\right)\delta Z = -\frac{g}{\alpha_T c_1^2}\delta Z$$

....(26)

$$Or \ \ \delta T_0 = -\delta\left(\frac{gZ}{\alpha_T c_1^2}\right) = 0 \Rightarrow \delta\left[T_0 + \frac{gZ}{\alpha_T c_1^2}\right] = 0$$

Thus the quantity $\left[T_0 + \frac{gZ}{\alpha_T c_1^2}\right]$ is conserved in a stratified isohaline fluid. Define

$$\theta = T_0 + \frac{gZ}{\alpha_T c_1^2}$$

....(27)

The new temperature θ thus defined by (27) is known as the *potential temperature* of the parcel, which takes into consideration the effect of compressibility/expansion of a sinking/rising parcel. Analogously, the *potential density* ρ_θ of a parcel could be defined. The potential temperature θ and potential density ρ_θ are simply the temperature and density that a parcel of seawater would have, if it were lifted adiabatically from its initial depth z to the reference level (ocean surface). Note that eqn. (26) has led to the definition of an important invariant θ for the ocean dynamics. A rigorous derivation could be arrived at from the thermodynamic equation. It may be noted that salinity, like mass, is also conserved. From eqn. (26), it may be noted that the parcel temperature T_o is approximately conserved if the speed of sound (c_1) is large. Also the variability of α_T and β could be neglected in the calculations. Indeed, ocean pressures are sufficiently high at depths to compress water parcels if taken adiabatically from surface to deeper levels. Moreover, adiabatic (without exchange of energy) compression will lead to *slight increase in temperature* of the sinking parcels therefore in defining the potential temperature, the effect of compression is removed.

Mixing of Parcels at Different Temperature and Salinity

In a process which takes place at constant pressure (dp=0) and density (dρ=0), the equation of state (2) gives the slope of the isopycnal relation T=T(S) as

$$\frac{dT}{dS} = \frac{\beta}{\alpha_T}$$ ....(28)

(28). In pure water, $\alpha_{T\,=}\,0$ at 4°C and increases with temperature. In the entire range of salinity of seawater α_T increases because the temperature of maximum density (which is 4°C) is depressed by the salt content. Consequently the slope of *T-S* curves given by the eqn. (28) will increase as temperature decreases.

Suppose m gm of water at T_1 and S_2 mixes thoroughly with (1-m) gm of water at T_2 and S_2 as shown in Figure. The resultant temperature and salinity of the mixture are then given by

$$\bar{T} = mT_1 + (1-m)T_2 \qquad (Conversation\ of\ int\ ernal\ energy)$$
$$\bar{S} = mS_1 + (1-m)S_2 \qquad (Conversation\ of\ salt\ content)$$

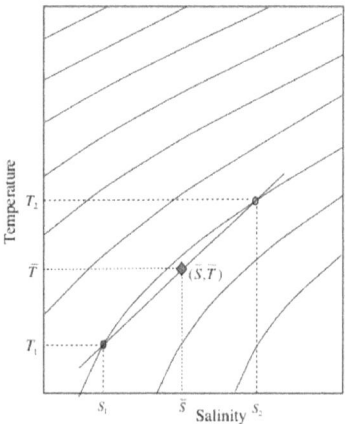

Mixing of two parcels at different temperatures and salinities: The mixed waters are heavier than the parcels. The illustration here takes the parcels at the same density which on mixing produce dense waters. However, the parcels at same density hardly mix, but it could happen under some external forcing arsing from bottom topography or mechanical forcing of wind at the sea surface.

Also note that slopes of the isopicnal curves increase with decreasing temperatures. The mean temperature and salinity after mixing are calculated.

$$\bar{T} = \frac{T_1 + T_2}{2} \ and\ \bar{S} = \frac{S_1 + S_2}{2}$$

The mixture at T and S lies on the straight line connecting the points (T_1, S_1) and (T_2, S_2) in the *T-S* diagram and its density is greater than the density of the original components. Since the mixture at point (S, T) is heavier than the original state of the parcels on the *T-S* curve, it has dynamical consequences on the stability of uniform density fluid with stratification in equilibrium. A small perturbation to this equilibrium will cause some mixing and the mixture will have greater density than the basic state. The fluid mixture so formed will sink and thereby generate new motions and additional mixing. Which means, "the basic state of uniform density is *unstable* because of variable α_T". This is the *"caballing effect"*, which arises due to variation in α_T at low temperatures in the ocean. Thus denser waters are formed through caballing effect when waters of different temperature and salinity mix. The currents associated with caballing have thermohaline circulation pattern. The caballing effect could be forced by the action winds on the surface waters.

When horizontal mixing takes place in the ocean, it most effectively happens along surfaces of constant σ_t since buoyant force does not inhibit exchange. As reasoned above and explained in Figure, the mixed waters will tend to sink below the unmixed waters (the caballing effect). Indeed *T-S* characteristics continuously vary over the ocean and it is very unlikely that waters of same σ_t but of very different temperature and salinity might mix directly. In other words, vertical circulations caused by the caballing effect may not be important.

The Density Parameter

The density anomaly σ_t as defined in eqn. (13) is the most widely used parameter in ocean science. The *potential density* is that density seawater parcel which it would have, if it were brought adiabatically to the surface (the reference level). Since its temperature would then be θ, the potential temperature, the potential density is written as $\rho(0, \theta, S)$ and density anomaly is defined as

$$\sigma_\theta = \rho(0, \theta, S) - 1000(kg/m^3)$$ ....(29)

That is, sigma- θ is also defined just as $\sigma_t = [\rho(p, T, S) - 1000]$ as in (13). Moreover, the pressure effect has been removed in both σ_θ and σ_t. However the difference between these two quantities is less than $0.2\sigma_t$ for a 6000 *m* change in the depth. For this reason the density anomaly parameter σ_t is loosely referred to as a *potential density function*. Note that *in situ* increase in density is almost linear with depth, the average value of σ_t lies in the range 36 - 37kg/m_3, i.e. mean density of ocean is 1036-1037 kg/m 3. Nevertheless, the nonlinearity of the equation of state becomes important at deeper depths.

One can define the stability parameter N^2 in terms of the potential density as

$$N^2 = \frac{g}{\rho}\frac{\partial \sigma_t}{\partial Z} \quad or \quad N^2 = \frac{g}{\rho}\frac{\partial \sigma_\theta}{\partial Z} \quad or \quad N^2 = \frac{g}{\rho}\left[\frac{\partial \rho}{\partial S} + \frac{\partial S}{\partial Z}\left(\frac{dT}{dZ} - \Gamma\right)\right]$$ ....(30)

Where Γ is the increase in temperature per unit increase in depth under adiabatic conditions. The results so far established for the seawater are summarized in Table: given below.

Table: Dependence of σ_t, α_T and β on temperature and salinity at three typical points, viz., a, b and c

Depth	(a)	(b)	(c)	Inference from T-S plots
Surface				
$T_o(°C)$	-1.5	5	15	
$\alpha_T(x10^{-4}\ K^{-1})$	0.3	1	2	1. Salty water is denser than fresh water; warm water is less dense than cold water.
$S_o(psu)$	34	36	38	
$\beta(x\ 10^{-4}\ psu^{-1})$	7.8	7.8	7.6	2. Fresh water (S=0) is heavier at 4°C i.e. has maximum density; fresh water colder than this is less dense. Hence ice forms on the top of fresh water lakes. Cooling from surface in winter forms ice rather than denser water.
$\sigma_t(kg/m^3)$	28	29	28	
1 km deep				3. σ_t varies monotonically with temperature. At sea surface σ_t = 26 kg/m³ (typical value). In open ocean temperatures influence density. (iv) Potential temperature and potential density are defined in analogy with atmosphere allowing for compressibility effects; sea surface is the reference level.
$T_o(°C)$	-1.5	3	13	
$\alpha_T(x\ 10^{-4}\ K^{-1})$	0.65	1.1	2.2	
$S_o(psu)$	34	35	38	
$\beta(x\ 10^{-4}\ psu^{-1})$	7.1	7.7	7.4	
$\sigma_t(kg/m^3)$	-3	0.6	6.9	Density of seawater (accuracy required): 10^{-5}

First Law of Thermodynamics for Seawater

Thermodynamics is about energy exchange within systems, between systems and their surroundings. It connects the observable with general principles, and its prime goal is to determine of thermodynamical potentials (such as internal energy, entropy). Seawater is considered as a solution consisting of a single solute; and sea salt is thus considered a lumped quantity representing different dissolved constituents in ocean waters. The mean molecular weight of sea salt is 62.8 g/mole, and its mean ionic weight is 31.4 g/mole.

The first law of thermodynamics is an assertion of conservation of energy, particularly the internal energy of the system. The internal energy is the store of energy in the system at equilibrium, and it has the same value no matter how equilibrium has been attained. The working substance is the seawater in all discussions of ocean thermodynamics. In accordance to the lexicon of thermodynamics, e is a state variable that depends upon the independent state variables S (Salinity), T (temperature) and p (pressure). If de is the change in internal energy per unit mass of the working substance due to the change dq in the heat content per unit mass and dw amount of the work done on the system, then the first law of thermodynamics states that

$$de = dq + dw \text{ (J kg}^{-1}) \dots\dots\dots\dots\dots\dots\dots\dots(31)$$

Mathematically, de is an exact differential (or perfect differential) while dq and dw are not exact differentials as they are not state variables. Since S, T and p are independent state variables, de depends only on the values of these variables and not on the history of the transformations that led to this state; that is why, de is a perfect differential.

In eqn. (31), dq is the amount of heat flowing into the system and dw is amount of work performed on the system. Thus if $d\alpha$ is change in the specific volume (i.e. volume per unit mass), then

$$dw = -p \, d\alpha \qquad\qquad\qquad\qquad\qquad\qquad\qquad\qquad (32)$$

If different chemical substances are added to a sample of water, then the internal energy of the system will change in accordance to the chemical potential of each substance. If dn_i moles of a substance with chemical potential μ_i are present in a sample of unit mass, then the internal energy of the seawater is expressed as

$$de = dq - pd\alpha + \sum_i \mu_i dn_i + L_v dn \qquad\qquad\qquad\qquad\qquad\qquad (33)$$

$\sum_i \mu_i dn_i$ is the change in de due to different chemical constituents; $L_v dn$ is change that comes from evaporation which is present at all temperatures from ocean surface. L_v is latent heat vaporization per mole and dn is the number of moles evaporated. Except the term dq, each term is a product of an *intensive variable* (pressure, chemical potential, etc.) and size of the system (α, n_i etc.). The application of second law of thermodynamics removes this lack of symmetry arising due to the presence of dq in the expression (33). A thermodynamical variable or property that is independent of the size of the system is said to be an *intensive variable*. The variables, which depend on the size of the system, are called the *extensive variables* and are therefore first order homogeneous functions.

In the formulation (33), the kinetic energy and potential energy of the fluid flow have not been considered. Though these mechanical forms of energy could be converted to thermodynamical forms through friction and other dissipative forces, but in the sea these changes are small compared to other energy inputs. The chemical potentials could be written in terms of a combined chemical potential of all salts in the seawater. Hence the change in the internal energy due to dS change relative of pure water is given by

$$\mu dS = \sum_i \mu_i dn_i \qquad\qquad\qquad\qquad\qquad\qquad\qquad\qquad (34)$$

S is the total salinity of seawater. For fluids, the enthalpy per unit mass, h, is a convenient thermodynamic potential defined as

$$h = e + p\alpha \qquad\qquad\qquad\qquad\qquad\qquad\qquad\qquad\qquad (35)$$

The differentiation of (35) gives

$$dh = de + pd\alpha + \alpha \, dp \qquad\qquad\qquad\qquad\qquad\qquad\qquad (36)$$

Substituting for de from eqn. (33) and using (34), we get an expression for change of enthalpy as

$$dh = dq + \alpha \, dp + \mu \, dS + L_v dn \qquad\qquad\qquad\qquad\qquad\qquad (37)$$

The term αdp represents the pressure-volume change in (37). Moreover, for fluid systems, the required coefficients can be easily evaluated at constant pressure than at constant density (e.g. gases); hence, it is preferable to use enthalpy as a state variable.

Second Law of Thermodynamics

The second law of thermodynamics states that the total entropy of an isolated system can only increase over time or it can remain constant in ideal cases where the system is in a steady state (equilibrium) or undergoing a reversible process. The increase in entropy accounts for the irreversibility of natural processes, and the asymmetry between future and past.

Historically, the second law was an empirical finding that was accepted as an axiom of thermodynamic theory. Statistical thermodynamics, classical or quantum, explains the microscopic origin of the law.

The second law has been expressed in many ways. Its first formulation is credited to the French scientist Sadi Carnot in 1824, who showed that there is an upper limit to the efficiency of conversion of heat to work in a heat engine.

Introduction

The first law of thermodynamics provides the basic definition of internal energy, associated with all thermodynamic systems, and states the rule of conservation of energy. The second law is concerned with the direction of natural processes. It asserts that a natural process runs only in one sense, and is not reversible. For example, heat always flows spontaneously from hotter to colder bodies, and never the reverse, unless external work is performed on the system. Its modern definition is in terms of entropy.

In a fictive reversible process, an infinitesimal increment in the entropy (dS) of a system is defined to result from an infinitesimal transfer of heat (δQ) to a closed system divided by the common temperature (T) of the system and the surroundings which supply the heat:

$$dS = \frac{\delta Q}{T} \qquad \text{(closed system, idealized fictive reversible process).}$$

Different notations are used for infinitesimal amounts of heat (δ) and infinitesimal amounts of entropy (d) because entropy is a function of state, while heat, like work, is not. For an actually possible infinitesimal process without exchange of matter with the surroundings, the second law requires that the increment in system entropy be greater than that:

$$dS > \frac{\delta Q}{T} \qquad \text{(closed system, actually possible, irreversible process).}$$

This is because a general process for this case may include work being done on the system by its surroundings, which must have frictional or viscous effects inside the system, and because heat transfer actually occurs only irreversibly, driven by a finite temperature difference.

The zeroth law of thermodynamics in its usual short statement allows recognition that two bodies in a relation of thermal equilibrium have the same temperature, especially that a test body has the same temperature as a reference thermometric body. For a body in thermal equilibrium with another, there are indefinitely many empirical temperature scales, in general respectively depending on the properties of a particular reference thermometric body. The second law allows a distinguished temperature scale, which defines an absolute, thermodynamic temperature, independent of the properties of any particular reference thermometric body.

Various Statements of the Law

The second law of thermodynamics may be expressed in many specific ways, the most prominent classical statements being the statement by Rudolf Clausius (1854), the statement by Lord Kelvin (1851), and the statement in axiomatic thermodynamics by Constantin Carathéodory (1909). These statements cast the law in general physical terms citing the impossibility of certain processes. The Clausius and the Kelvin statements have been shown to be equivalent.

Carnot's Principle

The historical origin of the second law of thermodynamics was in Carnot's principle. It refers to a cycle of a Carnot heat engine, fictively operated in the limiting mode of extreme slowness known as quasi-static, so that the heat and work transfers are between subsystems that are always in their own internal states of thermodynamic equilibrium. The Carnot engine is an idealized device of special interest to engineers who are concerned with the efficiency of heat engines. Carnot's principle was recognized by Carnot at a time when the caloric theory of heat was seriously considered, before the recognition of the first law of thermodynamics, and before the mathematical expression of the concept of entropy. Interpreted in the light of the first law, it is physically equivalent to the second law of thermodynamics, and remains valid today. It states

The efficiency of a quasi-static or reversible Carnot cycle depends only on the temperatures of the two heat reservoirs, and is the same, whatever the working substance. A Carnot engine operated in this way is the most efficient possible heat engine using those two temperatures.

Clausius Statement

The German scientist Rudolf Clausius laid the foundation for the second law of thermodynamics in 1850 by examining the relation between heat transfer and work. His formulation of the second law, which was published in German in 1854, is known as the *Clausius statement*:

Heat can never pass from a colder to a warmer body without some other change, connected therewith, occurring at the same time.

The statement by Clausius uses the concept of 'passage of heat'. As is usual in thermodynamic discussions, this means 'net transfer of energy as heat', and does not refer to contributory transfers one way and the other.

Heat cannot spontaneously flow from cold regions to hot regions without external work being performed on the system, which is evident from ordinary experience of refrigeration, for example. In a refrigerator, heat flows from cold to hot, but only when forced by an external agent, the refrigeration system.

Kelvin Statement

Lord Kelvin expressed the second law as

It is impossible, by means of inanimate material agency, to derive mechanical effect from any portion of matter by cooling it below the temperature of the coldest of the surrounding objects.

Equivalence of the Clausius and the Kelvin statements

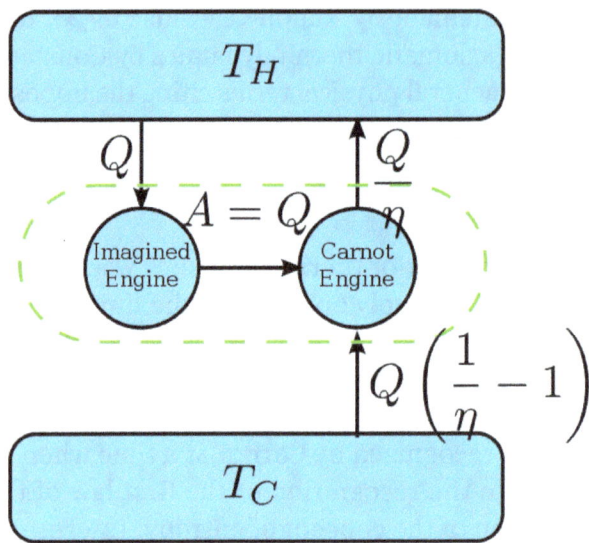

Derive Kelvin Statement from Clausius Statement

Suppose there is an engine violating the Kelvin statement: i.e., one that drains heat and converts it completely into work in a cyclic fashion without any other result. Now pair it with a reversed Carnot engine as shown by the figure. The net and sole effect of this newly created engine consisting of the two engines mentioned is transferring heat $\Delta Q = Q\left(\dfrac{1}{\eta} - 1\right)$ from the cooler reservoir to the hotter one, which violates the Clausius statement. Thus a violation of the Kelvin statement implies a violation of the Clausius statement, i.e. the Clausius statement implies the Kelvin statement. We can prove in a similar manner that the Kelvin statement implies the Clausius statement, and hence the two are equivalent.

Planck's Proposition

Planck offered the following proposition as derived directly from experience. This is sometimes regarded as his statement of the second law, but he regarded it as a starting point for the derivation of the second law.

> *It is impossible to construct an engine which will work in a complete cycle, and produce no effect except the raising of a weight and cooling of a heat reservoir.*

Relation between Kelvin's Statement and Planck's Proposition

It is almost customary in textbooks to speak of the "Kelvin-Planck statement" of the law, as for example in the text by ter Haar and Wergeland. One text gives a statement very like Planck's proposition, but attributes it to Kelvin without mention of Planck. One monograph quotes Planck's proposition as the "Kelvin-Planck" formulation, the text naming Kelvin as its author, though it correctly cites Planck in its references. The reader may compare the two statements quoted just above here.

Planck's Statement

Planck stated the second Law as follows.

> *Every process occurring in nature proceeds in the sense in which the sum of the entropies of all bodies taking part in the process is increased. In the limit, i.e. for reversible processes, the sum of the entropies remains unchanged.*

Rather like Planck's statement is that of Uhlenbeck and Ford for *irreversible phenomena.*

> ... in an irreversible or spontaneous change from one equilibrium state to another (as for example the equalization of temperature of two bodies A and B, when brought in contact) the entropy always increases.

Principle of Carathéodory

Constantin Carathéodory formulated thermodynamics on a purely mathematical axiomatic foundation. His statement of the second law is known as the Principle of Carathéodory, which may be formulated as follows:

In every neighborhood of any state S of an adiabatically enclosed system there are states inaccessible from S.

With this formulation, he described the concept of adiabatic accessibility for the first time and provided the foundation for a new subfield of classical thermodynamics, often called geometrical thermodynamics. It follows from Carathéodory's principle that quantity of energy quasi-statically transferred as heat is a holonomic process function, in other words, $\delta Q = TdS$.

Though it is almost customary in textbooks to say that Carathéodory's principle expresses the second law and to treat it as equivalent to the Clausius or to the Kelvin-Planck statements, such is not the case. To get all the content of the second law, Carathéodory's principle needs to be supplemented by Planck's principle, that isochoric work always increases the internal energy of a closed system that was initially in its own internal thermodynamic equilibrium.

Planck's Principle

In 1926, Max Planck wrote an important paper on the basics of thermodynamics. He indicated the principle

> The internal energy of a closed system is increased by an adiabatic process, throughout the duration of which, the volume of the system remains constant.

This formulation does not mention heat and does not mention temperature, nor even entropy, and does not necessarily implicitly rely on those concepts, but it implies the content of the second law. A closely related statement is that "Frictional pressure never does positive work." Using a now-obsolete form of words, Planck himself wrote: "The production of heat by friction is irreversible."

Not mentioning entropy, this principle of Planck is stated in physical terms. It is very closely related to the Kelvin statement given just above. It is relevant that for a system at constant volume and mole numbers, the entropy is a monotonic function of the internal energy. Nevertheless, this principle of Planck is not actually Planck's preferred statement of the second law, which is quoted above, in a previous sub-section of the present section of this present article, and relies on the concept of entropy.

A statement that in a sense is complementary to Planck's principle is made by Borgnakke and Sonntag. They do not offer it as a full statement of the second law:

> … there is only one way in which the entropy of a [closed] system can be decreased, and that is to transfer heat from the system.

Differing from Planck's just foregoing principle, this one is explicitly in terms of entropy change. Of course, removal of matter from a system can also decrease its entropy.

Statement for a System that has a Known Expression of its Internal Energy as a Function of its Extensive State Variables

The second law has been shown to be equivalent to the internal energy U being a weakly convex function, when written as a function of extensive properties (mass, volume, entropy, …).

Corollaries

Perpetual Motion of the Second Kind

Before the establishment of the Second Law, many people who were interested in inventing a perpetual motion machine had tried to circumvent the restrictions of first law of thermodynamics by extracting the massive internal energy of the environment as the power of the machine. Such a machine is called a "perpetual motion machine of the second kind". The second law declared the impossibility of such machines.

Carnot Theorem

Carnot's theorem (1824) is a principle that limits the maximum efficiency for any possible engine. The efficiency solely depends on the temperature difference between the hot and cold thermal reservoirs. Carnot's theorem states:

- All irreversible heat engines between two heat reservoirs are less efficient than a Carnot engine operating between the same reservoirs.

- All reversible heat engines between two heat reservoirs are equally efficient with a Carnot engine operating between the same reservoirs.

In his ideal model, the heat of caloric converted into work could be reinstated by reversing the motion of the cycle, a concept subsequently known as thermodynamic reversibility. Carnot, however, further postulated that some caloric is lost, not being converted to mechanical work. Hence, no real heat engine could realise the Carnot cycle's reversibility and was condemned to be less efficient.

Though formulated in terms of caloric, rather than entropy, this was an early insight into the second law.

Clausius Inequality

The Clausius theorem (1854) states that in a cyclic process

$$\oint \frac{\delta Q}{T} \leq 0.$$

The equality holds in the reversible case and the '<' is in the irreversible case. The reversible case is used to introduce the state function entropy. This is because in cyclic processes the variation of a state function is zero from state functionality.

Thermodynamic Temperature

For an arbitrary heat engine, the efficiency is:

$$\eta = \frac{W_n}{q_H} = \frac{q_H - q_C}{q_H} = 1 - \frac{q_C}{q_H} \qquad (1)$$

where W_n is for the net work done per cycle. Thus the efficiency depends only on q_C/q_H.

Carnot's theorem states that all reversible engines operating between the same h eat reservoirs are equally efficient. Thus, any reversible heat engine operating between temperatures T_1 and T_2 must have the same efficiency, that is to say, the efficiency is the function of temperatures only:

$$\frac{q_C}{q_H} = f(T_H, T_C) \qquad (2).$$

In addition, a reversible heat engine operating between temperatures T_1 and T_3 must have the same efficiency as one consisting of two cycles, one between T_1 and another (intermediate) temperature T_2, and the second between T_2 and T_3. This can only be the case if

$$f(T_1, T_3) = \frac{q_3}{q_1} = \frac{q_2 q_3}{q_1 q_2} = f(T_1, T_2) f(T_2, T_3).$$

Now consider the case where T_1 is a fixed reference temperature: the temperature of the triple point of water. Then for any T_2 and T_3,

$$f(T_2, T_3) = \frac{f(T_1, T_3)}{f(T_1, T_2)} = \frac{273.16 \cdot f(T_1, T_3)}{273.16 \cdot f(T_1, T_2)}.$$

Therefore, if thermodynamic temperature is defined by

$$T = 273.16 \cdot f(T_1, T)$$

then the function f, viewed as a function of thermodynamic temperature, is simply

$$f(T_2, T_3) = \frac{T_3}{T_2},$$

and the reference temperature T_1 will have the value 273.16. (Of course any reference temperature and any positive numerical value could be used—the choice here corresponds to the Kelvin scale.)

Entropy

According to the Clausius equality, for a reversible process

$$\oint \frac{\delta Q}{T} = 0$$

That means the line integral $\int_L \frac{\delta Q}{T}$ is path independent.

So we can define a state function S called entropy, which satisfies

$$dS = \frac{\delta Q}{T}$$

With this we can only obtain the difference of entropy by integrating the above formula. To obtain the absolute value, we need the Third Law of Thermodynamics, which states that S=0 at absolute zero for perfect crystals.

For any irreversible process, since entropy is a state function, we can always connect the initial and terminal states with an imaginary reversible process and integrating on that path to calculate the difference in entropy.

Now reverse the reversible process and combine it with the said irreversible process. Applying Clausius inequality on this loop,

$$-\Delta S + \int \frac{\delta Q}{T} = \oint \frac{\delta Q}{T} < 0$$

Thus,

$$\Delta S \geq \int \frac{\delta Q}{T}$$

where the equality holds if the transformation is reversible.

Notice that if the process is an adiabatic process, then $\delta Q = 0$, so $\Delta S \geq 0$.

Energy, Available Useful Work

An important and revealing idealized special case is to consider applying the Second Law to the scenario of an isolated system (called the total system or universe), made up of two parts: a sub-system of interest, and the sub-system's surroundings. These surroundings are imagined to be so large that they can be considered as an *unlimited* heat reservoir at temperature T_R and pressure P_R — so that no matter how much heat is transferred to (or from) the sub-system, the temperature of the surroundings will remain T_R; and no matter how much the volume of the sub-system expands (or contracts), the pressure of the surroundings will remain P_R.

Whatever changes to dS and dS_R occur in the entropies of the sub-system and the surroundings individually, according to the Second Law the entropy S_{tot} of the isolated total system must not decrease:

$$dS_{tot} = dS + dS_R \geq 0$$

According to the First Law of Thermodynamics, the change dU in the internal energy of the sub-system is the sum of the heat δq added to the sub-system, *less* any work δw done *by* the sub-system, *plus* any net chemical energy entering the sub-system $d \sum \mu_{iR} N_i$, so that:

$$dU = \delta q - \delta w + d(\sum \mu_{iR} N_i)$$

where μ_{iR} are the chemical potentials of chemical species in the external surroundings.

Now the heat leaving the reservoir and entering the sub-system is

$$\delta q = T_R(-dS_R) \leq T_R dS$$

where we have first used the definition of entropy in classical thermodynamics (alternatively, in statistical thermodynamics, the relation between entropy change, temperature and absorbed heat can be derived); and then the Second Law inequality from above.

It therefore follows that any net work δw done by the sub-system must obey

$$\delta w \leq -dU + T_R dS + \sum \mu_{iR} dN_i$$

It is useful to separate the work δw done by the subsystem into the *useful* work δw_u that can be done *by* the sub-system, over and beyond the work $p_R\, dV$ done merely by the sub-system expanding against the surrounding external pressure, giving the following relation for the useful work (exergy) that can be done:

$$\delta w_u \leq -d(U - T_R S + p_R V - \sum \mu_{iR} N_i)$$

It is convenient to define the right-hand-side as the exact derivative of a thermodynamic potential, called the *availability* or *exergy E* of the subsystem,

$$E = U - T_R S + p_R V - \sum \mu_{iR} N_i$$

The Second Law therefore implies that for any process which can be considered as divided simply into a subsystem, and an unlimited temperature and pressure reservoir with which it is in contact,

$$dE + \delta w_u \leq 0$$

i.e. the change in the subsystem's exergy plus the useful work done *by* the subsystem (or, the change in the subsystem's exergy less any work, additional to that done by the pressure reservoir, done *on* the system) must be less than or equal to zero.

In sum, if a proper *infinite-reservoir-like* reference state is chosen as the system surroundings in the real world, then the Second Law predicts a decrease in E for an irreversible process and no change for a reversible process.

$$dS_{tot} \geq 0 \text{ Is equivalent to } dE + \delta w_u \leq 0$$

This expression together with the associated reference state permits a design engineer working at the macroscopic scale (above the thermodynamic limit) to utilize the Second Law without directly measuring or considering entropy change in a total isolated system. Those changes have already been considered by the assumption that the system under consideration can reach equilibrium with the reference state without altering the reference state. An efficiency for a process or collection of processes that compares it to the reversible ideal may also be found.

This approach to the Second Law is widely utilized in engineering practice, environmental accounting, systems ecology, and other disciplines.

History

Nicolas Léonard Sadi Carnot in the traditional uniform of a student of the École Polytechnique.

The first theory of the conversion of heat into mechanical work is due to Nicolas Léonard Sadi Carnot in 1824. He was the first to realize correctly that the efficiency of this conversion depends on the difference of temperature between an engine and its environment.

Recognizing the significance of James Prescott Joule's work on the conservation of energy, Rudolf Clausius was the first to formulate the second law during 1850, in this form: heat does not flow *spontaneously* from cold to hot bodies. While common knowledge now, this was contrary to the caloric theory of heat popular at the time, which considered heat as a fluid. From there he was able to infer the principle of Sadi Carnot and the definition of entropy (1865).

Established during the 19th century, the Kelvin-Planck statement of the Second Law says, "It is impossible for any device that operates on a cycle to receive heat from a single reservoir and produce a net amount of work." This was shown to be equivalent to the statement of Clausius.

The ergodic hypothesis is also important for the Boltzmann approach. It says that, over long periods of time, the time spent in some region of the phase space of microstates with the same energy is proportional to the volume of this region, i.e. that all accessible microstates are equally probable over a long period of time. Equivalently, it says that time average and average over the statistical ensemble are the same.

There is a traditional doctrine, starting with Clausius, that entropy can be understood in terms of molecular 'disorder' within a macroscopic system. This doctrine is obsolescent.

Account Given by Clausius

Rudolf Clausius

In 1856, the German physicist Rudolf Clausius stated what he called the "second fundamental theorem in the mechanical theory of heat" in the following form:

$$\int \frac{\delta Q}{T} = -N$$

where Q is heat, T is temperature and N is the "equivalence-value" of all uncompensated transformations involved in a cyclical process. Later, in 1865, Clausius would come to define "equivalence-value" as entropy. On the heels of this definition, that same year, the most famous version of the second law was read in a presentation at the Philosophical Society of Zurich on April 24, in which, in the end of his presentation, Clausius concludes:

The entropy of the universe tends to a maximum.

This statement is the best-known phrasing of the second law. Because of the looseness of its language, e.g. universe, as well as lack of specific conditions, e.g. open, closed, or isolated, many people take this simple statement to mean that the second law of thermodynamics applies virtually to every subject imaginable. This, of course, is not true; this statement is only a simplified version of a more extended and precise description.

In terms of time variation, the mathematical statement of the second law for an isolated system undergoing an arbitrary transformation is:

$$\frac{dS}{dt} \geq 0$$

where

> S is the entropy of the system and

> t is time.

The equality sign applies after equilibration. An alternative way of formulating of the second law for isolated systems is:

$$\frac{dS}{dt} = \dot{S}_i \text{ with } \dot{S}_i \geq 0$$

with $\dot{S}_i$ the sum of the rate of entropy production by all processes inside the system. The advantage of this formulation is that it shows the effect of the entropy production. The rate of entropy production is a very important concept since it determines (limits) the efficiency of thermal machines. Multiplied with ambient temperature T_a it gives the so-called dissipated energy $P_{diss} = T_a \dot{S}_i$.

The expression of the second law for closed systems (so, allowing heat exchange and moving boundaries, but not exchange of matter) is:

$$\frac{dS}{dt} = \frac{\dot{Q}}{T} + \dot{S}_i \text{ with } \dot{S}_i \geq 0$$

Here

> $\dot{Q}$ is the heat flow into the system

> T is the temperature at the point where the heat enters the system.

The equality sign holds in the case that only reversible processes take place inside the system. If irreversible processes take place (which is the case in real systems in operation) the >-sign holds. If heat is supplied to the system at several places we have to take the algebraic sum of the corresponding terms.

For open systems (also allowing exchange of matter):

$$\frac{dS}{dt} = \frac{\dot{Q}}{T} + \dot{S} + \dot{S}_i \text{ with } \dot{S}_i \geq 0$$

Here $\dot{S}$ is the flow of entropy into the system associated with the flow of matter entering the system. It should not be confused with the time derivative of the entropy. If matter is supplied at several places we have to take the algebraic sum of these contributions.

Statistical Mechanics

Statistical mechanics gives an explanation for the second law by postulating that a material is composed of atoms and molecules which are in constant motion. A particular set of positions and velocities for each particle in the system is called a microstate of the system and because of the constant motion, the system is constantly changing its microstate. Statistical mechanics postulates that, in equilibrium, each microstate that the system might be in is equally likely to occur, and when this assumption is made, it leads directly to the conclusion that the second law must hold in a statistical sense. That is, the second law will hold on average, with a statistical variation on the order of $1/\sqrt{N}$ where N is the number of particles in the system. For everyday (macroscopic) situations, the probability that the second law will be violated is practically zero. However, for systems with a small number of particles, thermodynamic parameters, including the entropy, may show significant statistical deviations from that predicted by the second law. Classical thermodynamic theory does not deal with these statistical variations.

Derivation from Statistical Mechanics

Due to Loschmidt's paradox, derivations of the Second Law have to make an assumption regarding the past, namely that the system is uncorrelated at some time in the past; this allows for simple probabilistic treatment. This assumption is usually thought as a boundary condition, and thus the second Law is ultimately a consequence of the initial conditions somewhere in the past, probably at the beginning of the universe (the Big Bang), though other scenarios have also been suggested.

Given these assumptions, in statistical mechanics, the Second Law is not a postulate, rather it is a consequence of the fundamental postulate, also known as the equal prior probability postulate, so long as one is clear that simple probability arguments are applied only to the future, while for the past there are auxiliary sources of information which tell us that it was low entropy. The first part of the second law, which states that the entropy of a thermally isolated system can only increase, is a trivial consequence of the equal prior probability postulate, if we restrict the notion of the entropy to systems in thermal equilibrium. The entropy of an isolated system in thermal equilibrium containing an amount of energy of E is:

$$S = k_{\mathrm{B}} \ln \left[\Omega(E) \right]$$

where $\Omega(E)$ is the number of quantum states in a small interval between E and $E + \delta E$. Here δE is a macroscopically small energy interval that is kept fixed. Strictly speaking this means that the entropy depends on the choice of δE. However, in the thermodynamic limit (i.e. in the limit of infinitely large system size), the specific entropy (entropy per unit volume or per unit mass) does not depend on δE.

Suppose we have an isolated system whose macroscopic state is specified by a number of variables. These macroscopic variables can, e.g., refer to the total volume, the positions of pistons in the system, etc. Then Ω will depend on the values of these variables. If a variable is not fixed, (e.g. we do not clamp a piston in a certain position), then because all the accessible states are equally likely in equilibrium, the free variable in equilibrium will be such that Ω is maximized as that is the most probable situation in equilibrium.

If the variable was initially fixed to some value then upon release and when the new equilibrium has been reached, the fact the variable will adjust itself so that Ω is maximized, implies that the entropy will have increased or it will have stayed the same (if the value at which the variable was fixed happened to be the equilibrium value). Suppose we start from an equilibrium situation and we suddenly remove a constraint on a variable. Then right after we do this, there are a number Ω of accessible microstates, but equilibrium has not yet been reached, so the actual probabilities of the system being in some accessible state are not yet equal to the prior probability of $1/\Omega$. We have already seen that in the final equilibrium state, the entropy will have increased or have stayed the same relative to the previous equilibrium state. Boltzmann's H-theorem, however, proves that the quantity H increases monotonically as a function of time during the intermediate out of equilibrium state.

Derivation of the Entropy Change for Reversible Processes

The second part of the Second Law states that the entropy change of a system undergoing a reversible process is given by:

$$dS = \frac{\delta Q}{T}$$

where the temperature is defined as:

$$\frac{1}{k_B T} \equiv \beta \equiv \frac{d \ln\left[\Omega(E)\right]}{dE}$$

See here for the justification for this definition. Suppose that the system has some external parameter, x, that can be changed. In general, the energy eigenstates of the system will depend on x. According to the adiabatic theorem of quantum mechanics, in the limit of an infinitely slow change of the system's Hamiltonian, the system will stay in the same energy eigenstate and thus change its energy according to the change in energy of the energy eigenstate it is in.

The generalized force, X, corresponding to the external variable x is defined such that $X dx$ is the work performed by the system if x is increased by an amount dx. E.g., if x is the volume, then X is the pressure. The generalized force for a system known to be in energy eigenstate E_r is given by:

$$X = -\frac{dE_r}{dx}$$

Since the system can be in any energy eigenstate within an interval of δE, we define the generalized force for the system as the expectation value of the above expression:

$$X = -\left\langle \frac{dE_r}{dx} \right\rangle$$

To evaluate the average, we partition the $\Omega(E)$ energy eigenstates by counting how many of them have a value for $\frac{dE_r}{dx}$ within a range between Y and $Y + \delta Y$. Calling this number $\Omega_Y(E)$, we have:

$$\Omega(E) = \sum_Y \Omega_Y(E)$$

The average defining the generalized force can now be written:

$$X = -\frac{1}{\Omega(E)} \sum_Y Y \Omega_Y(E)$$

We can relate this to the derivative of the entropy with respect to x at constant energy E as follows. Suppose we change x to x + dx. Then $\Omega(E)$ will change because the energy eigenstates depend on x, causing energy eigenstates to move into or out of the range between E and $E + \delta E$. Let's focus again on the energy eigenstates for which $\dfrac{dE_r}{dx}$ lies within the range between Y and $Y + \delta Y$. Since these energy eigenstates increase in energy by Y dx, all such energy eigenstates that are in the interval ranging from E − Y dx to E move from below E to above E. There are

$$N_Y(E) = \frac{\Omega_Y(E)}{\delta E} Y dx$$

such energy eigenstates. If $Y dx \leq \delta E$, all these energy eigenstates will move into the range between E and $E + \delta E$ and contribute to an increase in Ω. The number of energy eigenstates that move from below $E + \delta E$ to above $E + \delta E$ is, of course, given by $N_Y(E + \delta E)$. The difference

$$N_Y(E) - N_Y(E + \delta E)$$

is thus the net contribution to the increase in Ω. Note that if Y dx is larger than δE there will be the energy eigenstates that move from below E to above $E + \delta E$. They are counted in both $N_Y(E)$ and $N_Y(E + \delta E)$, therefore the above expression is also valid in that case.

Expressing the above expression as a derivative with respect to E and summing over Y yields the expression:

$$\left(\frac{\partial \Omega}{\partial x}\right)_E = -\sum_Y Y\left(\frac{\partial \Omega_Y}{\partial E}\right)_x = \left(\frac{\partial(\Omega X)}{\partial E}\right)_x$$

The logarithmic derivative of Ω with respect to x is thus given by:

$$\left(\frac{\partial \ln(\Omega)}{\partial x}\right)_E = \beta X + \left(\frac{\partial X}{\partial E}\right)_x$$

The first term is intensive, i.e. it does not scale with system size. In contrast, the last term scales as the inverse system size and will thus vanishes in the thermodynamic limit. We have thus found that:

$$\left(\frac{\partial S}{\partial x}\right)_E = \frac{X}{T}$$

Combining this with

$$\left(\frac{\partial S}{\partial E} \right)_x = \frac{1}{T}$$

Gives:

$$dS = \left(\frac{\partial S}{\partial E} \right)_x dE + \left(\frac{\partial S}{\partial x} \right)_E dx = \frac{dE}{T} + \frac{X}{T} dx = \frac{\delta Q}{T}$$

Derivation for Systems Described by the Canonical Ensemble

If a system is in thermal contact with a heat bath at some temperature T then, in equilibrium, the probability distribution over the energy eigenvalues are given by the canonical ensemble:

$$P_j = \frac{\exp\left(-\frac{E_j}{k_B T} \right)}{Z}$$

Here Z is a factor that normalizes the sum of all the probabilities to 1, this function is known as the partition function. We now consider an infinitesimal reversible change in the temperature and in the external parameters on which the energy levels depend. It follows from the general formula for the entropy:

$$S = -k_B \sum_j P_j \ln(P_j)$$

that

$$dS = -k_B \sum_j \ln(P_j) dP_j$$

Inserting the formula for P_j for the canonical ensemble in here gives:

$$dS = \frac{1}{T} \sum_j E_j dP_j = \frac{1}{T} \sum_j d(E_j P_j) - \frac{1}{T} \sum_j P_j dE_j = \frac{dE + \delta W}{T} = \frac{\delta Q}{T}$$

Living Organisms

There are two principal ways of formulating thermodynamics, (a) through passages from one state of thermodynamic equilibrium to another, and (b) through cyclic processes, by which the system is left unchanged, while the total entropy of the surroundings is increased. These two ways help to understand the processes of life. This topic is mostly beyond the scope of this present article, but has been considered by several authors, such as Erwin Schrödinger, Léon Brillouin and Isaac Asimov. It is also the topic of current research.

To a fair approximation, living organisms may be considered as examples of (b). Approximately, an animal's physical state cycles by the day, leaving the animal nearly unchanged. Animals take in food, water, and oxygen, and, as a result of metabolism, give out breakdown products and heat. Plants take in radiative energy from the sun, which may be regarded as heat, and carbon dioxide and water. They give out oxygen. In this way they grow. Eventually they die, and their remains rot. This can be regarded as a cyclic process. Overall, the sunlight is from a high temperature source, the sun, and its energy is passed to a lower temperature sink, the soil. This is an increase of entropy of the surroundings of the plant. Thus animals and plants obey the second law of thermodynamics, considered in terms of cyclic processes. Simple concepts of efficiency of heat engines are hardly applicable to this problem because they assume closed systems.

From the thermodynamic viewpoint that considers (a), passages from one equilibrium state to another, only a roughly approximate picture appears, because living organisms are never in states of thermodynamic equilibrium. Living organisms must often be considered as open systems, because they take in nutrients and give out waste products. Thermodynamics of open systems is currently often considered in terms of passages from one state of thermodynamic equilibrium to another, or in terms of flows in the approximation of local thermodynamic equilibrium. The problem for living organisms may be further simplified by the approximation of assuming a steady state with unchanging flows. General principles of entropy production for such approximations are subject to unsettled current debate or research. Nevertheless, ideas derived from this viewpoint on the second law of thermodynamics are enlightening about living creatures.

Gravitational Systems

In systems that do not require for their descriptions the general theory of relativity, bodies always have positive heat capacity, meaning that the temperature rises with energy. Therefore, when energy flows from a high-temperature object to a low-temperature object, the source temperature is decreased while the sink temperature is increased; hence temperature differences tend to diminish over time. This is not always the case for systems in which the gravitational force is important and the general theory of relativity is required. Such systems can spontaneously change towards uneven spread of mass and energy. This applies to the universe in large scale, and consequently it may be difficult or impossible to apply the second law to it. Beyond this, the thermodynamics of systems described by the general theory of relativity is beyond the scope of the present article.

Non-equilibrium States

The theory of classical or equilibrium thermodynamics is idealized. A main postulate or assumption, often not even explicitly stated, is the existence of systems in their own internal states of thermodynamic equilibrium. In general, a region of space containing a physical system at a given time, that may be found in nature, is not in thermodynamic equilibrium, read in the most stringent terms. In looser terms, nothing in the entire universe is or has ever been truly in exact thermodynamic equilibrium.

For purposes of physical analysis, it is often enough convenient to make an assumption of thermodynamic equilibrium. Such an assumption may rely on trial and error for its justification. If the assumption is justified, it can often be very valuable and useful because it makes available the theory of thermodynamics. Elements of the equilibrium assumption are that a system is observed to be unchanging over an indefinitely long time, and that there are so many particles in a system, that its particulate nature can be entirely ignored. Under such an equilibrium assumption, in general, there are no macroscopically detectable fluctuations. There is an exception, the case of critical states, which exhibit to the naked eye the phenomenon of critical opalescence. For laboratory studies of critical states, exceptionally long observation times are needed.

In all cases, the assumption of thermodynamic equilibrium, once made, implies as a consequence that no putative candidate "fluctuation" alters the entropy of the system.

It can easily happen that a physical system exhibits internal macroscopic changes that are fast enough to invalidate the assumption of the constancy of the entropy. Or that a physical system has so few particles that the particulate nature is manifest in observable fluctuations. Then the assumption of thermodynamic equilibrium is to be abandoned. There is no unqualified general definition of entropy for non-equilibrium states.

There are intermediate cases, in which the assumption of local thermodynamic equilibrium is a very good approximation, but strictly speaking it is still an approximation, not theoretically ideal. For non-equilibrium situations in general, it may be useful to consider statistical mechanical definitions of other quantities that may be conveniently called 'entropy', but they should not be confused or conflated with thermodynamic entropy properly defined for the second law. These other quantities indeed belong to statistical mechanics, not to thermodynamics, the primary realm of the second law.

The physics of macroscopically observable fluctuations is beyond the scope of this article.

Arrow of Time

The second law of thermodynamics is a physical law that is not symmetric to reversal of the time direction.

The second law has been proposed to supply an explanation of the difference between moving forward and backwards in time, such as why the cause precedes the effect (the causal arrow of time).

Irreversibility

Irreversibility in thermodynamic processes is a consequence of the asymmetric character of thermodynamic operations, and not of any internally irreversible microscopic properties of the bodies. Thermodynamic operations are macroscopic external interventions imposed on the participating bodies, not derived from their internal properties. There are reputed "paradoxes" that arise from failure to recognize this.

Loschmidt's Paradox

Loschmidt's paradox, also known as the reversibility paradox, is the objection that it should not be possible to deduce an irreversible process from the time-symmetric dynamics that describe the microscopic evolution of a macroscopic system.

In the opinion of Schrödinger, "It is now quite obvious in what manner you have to reformulate the law of entropy—or for that matter, all other irreversible statements—so that they be capable of being derived from reversible models. You must not speak of one isolated system but at least of two, which you may for the moment consider isolated from the rest of the world, but not always from each other." The two systems are isolated from each other by the wall, until it is removed by the thermodynamic operation, as envisaged by the law. The thermodynamic operation is externally imposed, not subject to the reversible microscopic dynamical laws that govern the constituents of the systems. It is the cause of the irreversibility. The statement of the law in this present article complies with Schrödinger's advice. The cause–effect relation is logically prior to the second law, not derived from it.

Poincaré Recurrence Theorem

The Poincaré recurrence theorem considers a theoretical microscopic description of an isolated physical system. This may be considered as a model of a thermodynamic system after a thermodynamic operation has removed an internal wall. The system will, after a sufficiently long time, return to a microscopically defined state very close to the initial one. The Poincaré recurrence time is the length of time elapsed until the return. It is exceedingly long, likely longer than the life of the universe, and depends sensitively on the geometry of the wall that was removed by the thermodynamic operation. The recurrence theorem may be perceived as apparently contradicting the second law of thermodynamics. More obviously, however, it is simply a microscopic model of thermodynamic equilibrium in an isolated system formed by removal of a wall between two systems. For a typical thermodynamical system, the recurrence time is so large (many many times longer than the lifetime of the universe) that, for all practical purposes, one cannot observe the recurrence. One might wish, nevertheless, to imagine that one could wait for the Poincaré recurrence, and then re-insert the wall that was removed by the thermodynamic operation. It is then evident that the appearance of irreversibility is due to the utter unpredictability of the Poincaré recurrence given only that the initial state was one of thermodynamic equilibrium, as is the case in macroscopic thermodynamics. Even if one could wait for it, one has no practical possibility of picking the right instant at which to re-insert the wall. The Poincaré recurrence theorem provides a solution to Loschmidt's paradox. If an isolated thermodynamic system could be monitored over increasingly many multiples of the average Poincaré recurrence time, the thermodynamic behavior of the system would become invariant under time reversal.

Maxwell's Demon

James Clerk Maxwell imagined one container divided into two parts, A and B. Both parts are filled with the same gas at equal temperatures and placed next to each other, separated by a wall. Observing the molecules on both sides, an imaginary demon guards a microscopic trapdoor in the wall. When a faster-than-average molecule from A flies towards the trapdoor, the demon opens it, and the molecule will fly from A to B. The average speed of the molecules in B will have increased while in A they will have slowed down on average. Since average molecular speed corresponds to temperature, the temperature decreases in A and increases in B, contrary to the second law of thermodynamics.

One response to this question was suggested in 1929 by Leó Szilárd and later by Léon Brillouin. Szilárd pointed out that a real-life Maxwell's demon would need to have some means of measuring molecular speed, and that the act of acquiring information would require an expenditure of energy.

Maxwell's demon repeatedly alters the permeability of the wall between A and B. It is therefore performing thermodynamic operations on a microscopic scale, not just observing ordinary spontaneous or natural macroscopic thermodynamic processes.

James Clerk Maxwell

Quotations

The law that entropy always increases holds, I think, the supreme position among the laws of Nature. If someone points out to you that your pet theory of the universe is in disagreement with Maxwell's equations — then so much the worse for Maxwell's equations. If it is found to be contradicted by observation — well, these experimentalists do bungle things sometimes. But if your theory is found to be against the second law of thermodynamics I can give you no hope; there is nothing for it but to collapse in deepest humiliation.

— Sir Arthur Stanley Eddington, The Nature of the Physical World (1927)

There have been nearly as many formulations of the second law as there have been discussions of it.

— Philosopher / Physicist P.W. Bridgman, (1941)

Clausius is the author of the sibyllic utterance, "The energy of the universe is constant; the entropy of the universe tends to a maximum." The objectives of continuum thermomechanics stop far short of explaining the "universe", but within that theory we may easily derive an explicit statement in some ways reminiscent of Clausius, but referring only to a modest object: an isolated body of finite size.

— Truesdell, C., Muncaster, R.G. (1980). Fundamentals of Maxwell's Kinetic Theory of a Simple Monatomic Gas, Treated as a Branch of Rational Mechanics, Academic Press, New York, ISBN 0-12-701350-4, p.17.

In eqn. (6.33), the heat differential dq was not expressed as a product of an intensive variable and an extensive variable while other terms do appear as a product of an intensive and extensive variable. Furthermore, dq is not a perfect differential. However, the second law removes this deficiency, which allows dq to be expressed as a product of an intensive and an extensive variable. The second law of thermodynamics states that the change in entropy per unit mass ($d\eta$) is greater than or equal to dq/T when an increment of heat dq is added to the system at temperature T; that is

$$d\eta \geq dq / T \dots\dots\dots\dots\dots\dots\dots\dots\dots\dots\dots\dots\dots\dots\dots\dots(38)$$

The eqn. (38) applies both to irreversible and reversible processes, however the equality in (38) strictly holds only for reversible processes. Moreover, entropy is a measure of disorder possessed by any system, so an addition of heat dq would increase disorder in system, for sure. Even for a reversible process that undergoes a thermodynamic cycle to return to its initial state, the entropy will increase in each cycle as the second law asserts. This implies that calculations performed with equality sign in (6.38) would place a lower bound on the entropy and the quantities derived from it are bounded. In fluid motions, entropy is proportional the amplitude of motion. Since changes in the ocean are rather slow, hence eqn. (6.38) with the equality sign can be confidently used.

It is now required to combine the laws of thermodynamics. According to first law the change in internal energy and enthalpy is given by

$$de = dq - p \, d\alpha + \mu \, ds + L_v dn \dots\text{(internal energy)}\dots\dots\dots\dots\dots\dots(39a)$$

$$dh = dq + \alpha \, dp + \mu \, ds + L_v dn \dots\text{(enthalpy)}\dots\dots\dots\dots\dots\dots\dots(39b)$$

The Second Law, with equality sign in (6.38) reads

$$d\eta = dq/T \;=> dq = Td\eta \dots\dots\dots\dots\dots\dots\dots\dots\dots\dots\dots\dots\dots(40)$$

In view of (40), the following expressions for change in internal energy and enthalpy change from (39) are given as

$$de = T \, d\eta - p \, d\alpha + \mu \, ds + L_v dn \dots\dots\dots\dots\dots\dots\dots\dots\dots\dots(41)$$

$$dh = T \, d\eta + \alpha \, dp + \mu \, ds + L_v dn \dots\dots\dots\dots\dots\dots\dots\dots\dots\dots(42)$$

The expression (41) and (42) are known as the Gibbs relations. The specific heat of water is large; and for sea water, the specific heat at constant pressure C_p = 3986 J kg^{-1} K^{-1} or 3850 J kg^{-1} K^{-1} at T = 20°C, S = 35 psu, p = 1.024 x10^5 Pa; The latent heat of vapourization is also very large. With T in °C, the expression for L_v is as follows,

$$L_v = (2.501 - 0.0039 \, T \,) \times 10^6 = 2453000 \text{ J/kg at } 20°C \dots\dots\dots\dots\dots\dots\dots(43)$$

There two simple facts influence the entire ocean-atmosphere system both on short and long time scales affecting the thermal inertia of the climate system. In time-dependent problems of geophysical fluid dynamics, time derivatives are required to be evaluated. The time derivatives for fluid flow situations are the convective derivatives (also known as: derivative following motion; substantial derivative; material derivative). Thus from (42) we have

$$\frac{Dh}{Dt} = T\frac{Dn}{Dt} + \alpha\frac{Dp}{Dt} + \mu\frac{DS}{Dt} + L_v\frac{Dn}{Dt} \, ; \; \frac{D(.)}{\partial t} + V.\nabla(.) \dots\dots\dots\dots\dots(44)$$

Consider the enthalpy per unit mass, h, as function dependent only on entropy (η), pressure (p) and salinity (S), then the Gibbs Relation (42) becomes

$$dh = Td\eta + \alpha dp + \mu ds \qquad h = h(\eta, p, S) \dots\dots\dots\dots\dots\dots\dots\dots(45)$$

For small changes in enthalpy, one can write from (44), dh as,

$$dh = \left(\frac{\partial h}{\partial \eta}\right)_{ps} d\eta + \left(\frac{\partial h}{\partial p}\right)_{\eta s} + \left(\frac{\partial h}{\partial \eta}\right)_{p\eta} dS$$ ....(46)

Comparing (6.45) and (6.46), we get additional thermodynamic equations:

$$T = \left(\frac{\partial h}{\partial \eta}\right)_{ps} K; \quad a + \left(\frac{\partial h}{\partial p}\right)_{\eta s} m^3 kg^{-1}; \quad \mu = \left(\frac{\partial h}{\partial S}\right)_{\eta p} J$$ ....(47)

Thus, relevant quantities in ocean thermodynamics have been derived that are used in discussing the stability of ocean stratification and to explain the formation of deep waters.

Salinity and Temperature Variations from Ocean Observations

In discussing the thermodynamics of the seawater, the variations in temperature and salinity play a key role. Such changes in these quantities together determine the density and therefore the type of waters – warm, cold, saline or fresh – in different areas of global ocean basins. A network of observing systems has been put in place by coordinated efforts of different countries. The Argo floats are a major current observing system over the global oceans. The daily fluctuations in sea surface temperature (SST) and sea surface salinity (SSS) mainly arise due the surface mixing under the mechanical forcing of the winds. The Argo floats in the Indian Ocean are deployed and maintained by India and the Indian National Centre for Ocean Information Services (INCOIS), Hyderabad, collects, manages and analyses the time series of variety of data produced by Argo floats. The data are constantly updated with newer data added to the time series and are also distributed for pedagogical and research purposes. A higher number of Argo floats is deployed in the areas of strong variations in ocean currents, eddies, sea surface temperature and salinity. Since the changes in the ocean thermodynamics could be well understood from the monthly observations too, we present monthly variations of salinity and temperature in the Indian Ocean for typical months that are available from INCOIS.

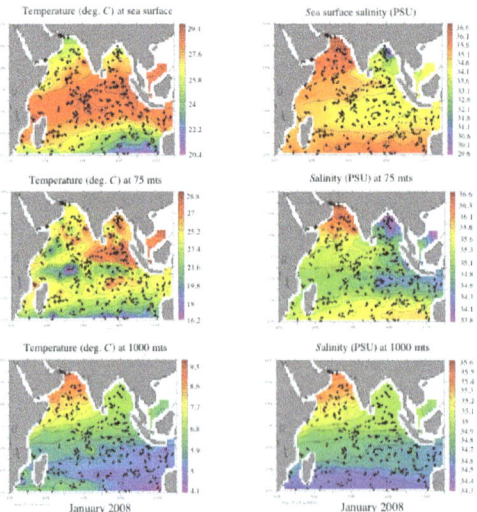

January 2008 temperature and salinity at depths 0 m (surface), 75 m, 1000 m in the Indian Ocean from Argo floats. The network of Argo floats to collect the data are represented by black dots in each panel.

The temperature (left panel) and salinity (right panel) for January 2008 in the top layer and at a depth of 1000 m in Figure show that both these variables portray strong latitudinal variations with higher SST in the equatorial region. However, in the upper (75 m depth) and lower (1000m depth) thermocline regions, temperatures are rather different with dominant zonal variations. Nevertheless, together with the latitudinal variations, the quasi-zonal uniformity in the salinity distribution may be noticed in Figure at the sea surface, 75 m and 1000 m in the Indian Ocean. The low salinity waters up to a depth of 75 m are present in the head Bay of Bengal due to river discharge, but due to excessive evaporation strong saline waters are found up to a depth of 1000 m in the Arabian Sea.

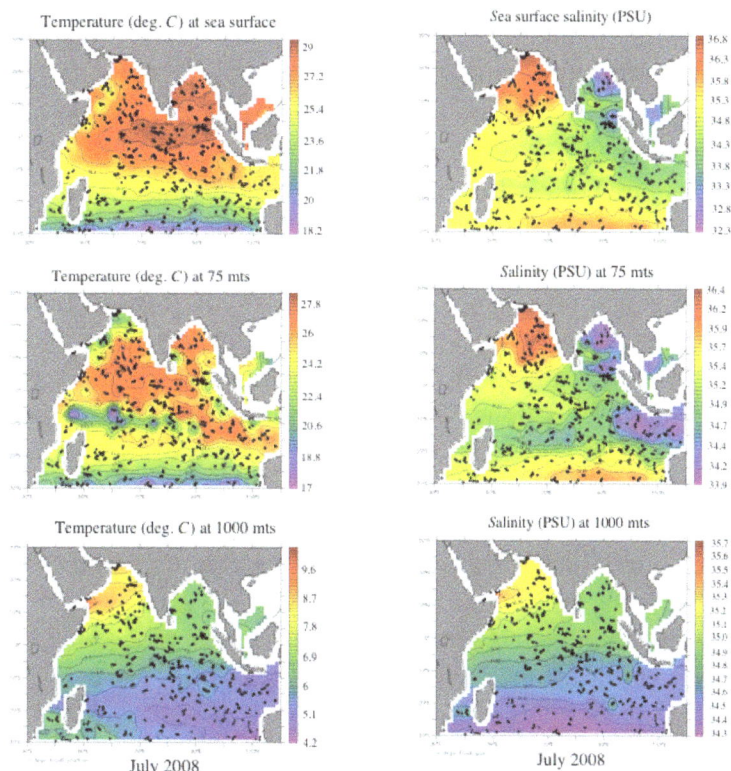

July 2008 temperature and salinity at depths o m (surface), 75 m, 1000 m in the Indian Ocean from Argo floats. The network of Argo floats to collect the data are represented by black dots in each panel.

The July distribution of temperature and salinity in the Indian Ocean during the year 2008 from Argo floats is presented in Figure at surface, 75 m and 1000 m. The key differences from January are noticed in temperatures as warm waters prevail upon both Arabian Sea and Bay of Bengal during July. Further, as the depth increases, a steeper temperature gradient may be noted in the Arabian Sea in contrast to Bay of Bengal during these typical months (January and July) of the year. However, over a greater part of the Indian Ocean at the depths around 1000 m, the water masses though practically have the same distribution of temperature and salinity yet some distinct variations are nonetheless noticeable in the southern part of the Indian Ocean. For example, significant thermodynamic variability may be noted south of Madagascar: temperatures around 6.5 °C and salinity 34.5 psu in January 2008 in comparison to 6.0 °C and 34.3 psu respectively in July. However the density of the water hardly change as a result of these variations, and vertical stratification is not disturbed.

Thus the density and the heat content of the ocean water are maintained despite changes in the temperature and salinity of oceanic water masses. From the thermal structure of the water columns in the Arabian Sea, there is an evidence of strong baroclinicity during the month of January with relatively colder waters at the surface and warm temperature at 1000 m depth. From the thermodynamic point of view the Bay of Bengal and the Arabian Sea thus appear to be quite different.

Adiabatic Effects in the Ocean

The density of ocean waters is determined mainly by temperature and salinity because pressure effects on density are sensibly negligible. The large scale circulation of ocean is slow and the water parcels which have origin in deep dense waters find their way to the surface of ocean through the path of thermohaline circulation. Cold and dense waters primarily reside in the high latitudes and lighter waters in the low latitudes. The variations of density also arise from depth wise lateral changes in temperature and salinity as evident from the equation of state. The surface waters move pole ward losing heat during their journey but their salinity increases due to excess evaporation. However, rainfall over the ocean decreases salinity of ocean waters and as shown in Figure, the salinity variations of seawaters can be directly explained in terms of the difference evaporation minus precipitation (E - P). The regions where evaporation exceeds precipitation, high salinity is obvious. That is why salinity peaks in mid latitudes where evaporation exceeds precipitation, but in the equatorial region low salinity waters are present because rainwater from deep convective clouds in the intertropical convergence zone (ITCZ) constantly mixes with the seawater. Even in the open sea, the surface freshwater and deeper cool high salinity waters would mix only when sufficiently strong winds force upwelling in such regions. Thus, ocean is in a state of mixing primarily due to surface wind forcing and several other processes; as a result, lapse rate in the ocean is constantly altered.

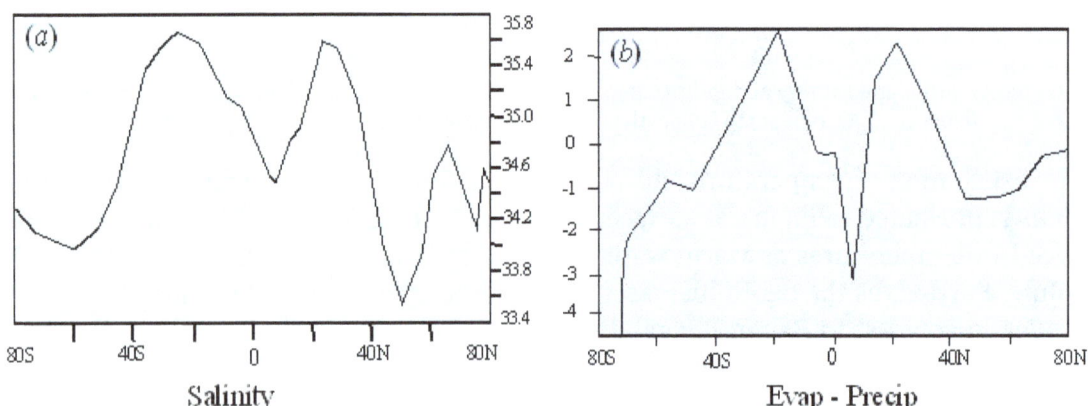

Salinity relation to evaporation minus precipitation.
(a) Zonally averaged salinity profile;
(b) Zonally averaged curve of difference of evaporation and precipitation.

We have presented in Figure, the density profile of ocean waters with depth which shows that seawater density rapidly increases in the top 50 m layer of the ocean. The steeper increase in density is noted from Figure up to a depth of 100 m.

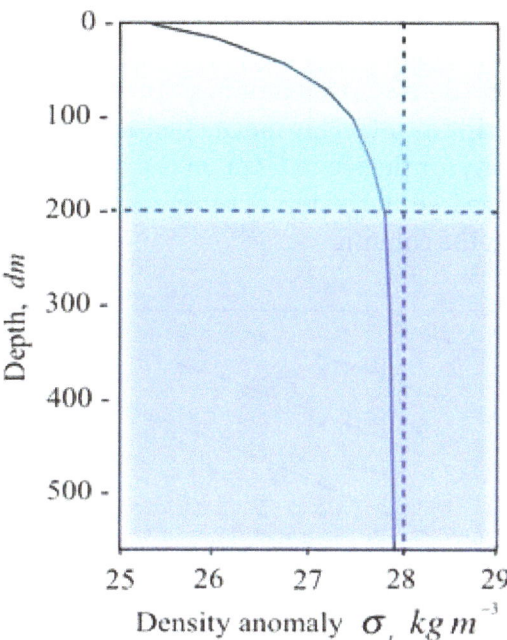

The density profile and water types in the upper oceanic layer.

(i) Lighter waters occupy the top 50 m of the ocean depth. Density rise with depth is steepest in this layer.

(ii) The rise in density is relatively slow with depth up to 200 m.

(iii) Below 200 m there is very slow rise in density with depth.

The density profile demonstrates that the top layer of 200 m profounder is very active and interacts and responds to the changes in the overlaying atmosphere

Figure presents a density profile on the salinity-temperature diagram obtained by plotting the observations at different depths in the Atlantic Ocean at 9°S starting from *150 m* to *5000 m*. On the salinity-temperature diagram, it is easier to interpret the types of Atlantic waters that occupy the body of the ocean depths. It means that the density profile is very instructive in understanding even the ocean circulation as the Atlantic Ocean spans from North Pole to South Pole. The waters moving along the ocean bottom reach from North Atlantic Ocean to Antarctica in several centuries (~2000 years). This kind of movement of water masses is associated with the thermohaline circulation that has been explained by Stommel and Aorons (1961). Now, let us analyse the density profile in the southern tropical Atlantic Ocean. The σ_t profile in Figure shows that at *150 m* warm high salinity waters are present, and the salinity constantly decreases with depth. The stratification in which higher density waters are arranged in deeper layer is mainly the result of colder temperatures. At a depth of *800 m,* the ocean temperatures are at 4°C (freshwater has maximum density at this temperature) and the density increases with depth as a result of salinity rise from *34.5 psu* at 800 m to *34.9 psu* at *1.6 km depth*. Then density of deeper waters from *1.4 km* to *2.0 km* increases due to colder temperatures and slight increase in salinity. However, most spectacular observation is the existence of two inflexion points in the profile. The density of seawater rises with the simultaneous fall of temperature and salinity and the first inflexion point occurs at 800 m as further rise in the density happens only due to rise in salinity from 34.5 *psu* and temperature remains practically constant at about 4°C. The second inflexion point occurs at 2000 m because in the deep ocean density remains almost constant though both temperature and salinity decrease up to a depth of 5000 m. Note that in fresh water, density is maximum at 4°C and it decrease if temperature further lowered with all the water freezing at 0°C. This means that there is only one inflexion point in the density of fresh water. Naturally, for the ice to form on the top of the sea surface there should be constant brine rejection from the freezing waters. The rejected brine adds to the lower level waters increasing the *in situ* density in the column.

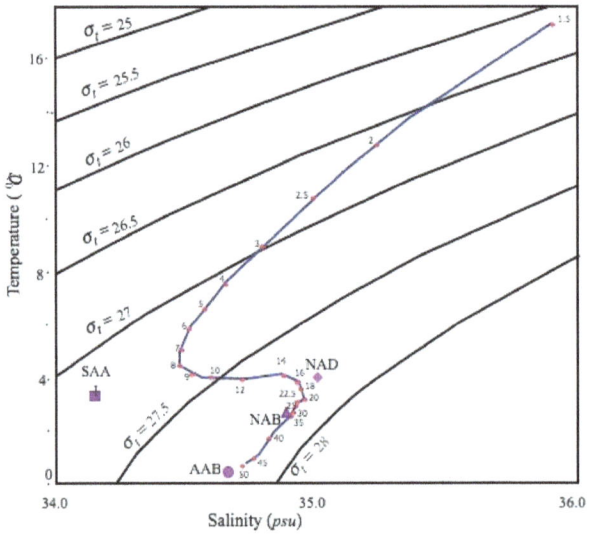

T-S diagram for Atlantic Ocean observation from 150 m to 5000 m at 9°S. Density increase from 150 m to 200 m is steepest. Two inflexion points in the density profile; first at 800 m where seawaters are at 4°C. The rise in density of Atlantic waters from 800 m to 1400 m is due to salinity increase and then due to the combined effect of temperature and salinity variations up to 2000 m. In the lowest layer from 2000 m to 5000 m the density of Atlantic waters is practically invariant.

(To get the depth in meters, multiply the number with 100; e.g. point marked 3 is at 3 x 100 = 300 m depth)

SAAI SubAntArctic Intermediate waters; NAD North Atlantic Deep waters

NAB North Atlantic Bottom waters; AAB AntArctic Bottom waters

Adiabatic Lapse Rate

What causes lapse rate in the sea? The adiabatic lapse rate is generally said to be proportional to the work done on a fluid parcel as its volume changes due to a pressure change (increase/ decrease). According to this definition, the adiabatic lapse rate Γ in the ocean would increase both with pressure and the fluid compressibility, but this is not the case. Indeed, Γ is proportional to thermal expansion coefficient and is independent of the compressibility of seawater. The adiabatic lapse rate Γ for seawater is defined as the rate at which *in situ* temperature (T) changes with pressure (p) while salinity (S) and entropy (η) are held constant, that is

$$\Gamma = \left(\frac{\partial T}{\partial p} \right)_{\eta s}$$ ....(1)

From the "fundamental thermodynamic relation (or Gibbs relation)" given by eqn. (45), we have the enthalpy change per unit mass,

$$dh = T\ d\eta + \alpha dp + \mu dS \quad \text{......................................(2)}$$

which takes the form

$$dh = C_p dT + \left(\frac{\partial h}{\partial p}\right)_{TS} dp + \left(\frac{\partial h}{\partial S}\right)_{Tp} dS \quad \text{......................................(3)}$$

Note that in writing eqn. (3), it has been assumed that

    h = h(T, p, S)

Hence, specific heat at constant pressure,

$$C_p = \left(\frac{\partial h}{\partial T}\right)_{pS} \quad \text{......................................(4)}$$

Also from equation (2), we have

$$T = \left(\frac{\partial h}{\partial \eta}\right)_{pS},\ \alpha = \left(\frac{\partial h}{\partial p}\right)_{S\eta},\ \mu = \left(\frac{\partial h}{\partial S}\right)_{p\eta} \quad \text{......................................(5)}$$

which on differentiation gives

$$\Gamma = \frac{\partial T}{\partial p} = \frac{\partial}{\partial p}\left[\frac{\partial h}{\partial \eta}\right|_{pS}\right] = \frac{\partial^2 h}{\partial p \partial \eta} = \frac{\partial^2 h}{\partial \eta \partial p} = \frac{\partial}{\partial \eta}\left(\frac{\partial h}{\partial p}\right)_{\eta s} = \left(\frac{\partial \alpha}{\partial \eta}\right)_{ps}\ \sin ce\left(\frac{\partial h}{\partial p}\right)_{\eta S} = \alpha.$$

So, we have

$$\Gamma = \left(\frac{\partial \alpha}{\partial \eta}\right)_{ps} = \frac{[\partial \alpha / \partial T]_{pS}}{[\partial \eta / \partial T]_{pS}} \quad \text{......................................(6)}$$

In (6), the expressions for $\left(\frac{\partial \alpha}{\partial T}\right)_{pS}$ and $\left(\frac{\partial \eta}{\partial T}\right)_{pS}$ and are to be obtained from the eqn. (3), which involve some algebra.

$$Now\quad C_p = \left(\frac{\partial h}{\partial T}\right)_{pS} \Rightarrow \quad C_p = \left(\frac{\partial h}{\partial \eta}\right)_{pS}\left(\frac{\partial \eta}{\partial T}\right)_{pS} \quad \text{......................................(7)}$$

$$C_p = T\left(\frac{\partial \eta}{\partial T}\right)_{pS} \Rightarrow \quad \left(\frac{\partial \eta}{\partial T}\right)_{pS} = \frac{C_p}{T} \quad \text{......................................(8)}$$

Hence, the expression for Γ given by (7.6) becomes

$$\Gamma = \frac{T}{C_p}\left(\frac{\partial \alpha}{\partial T}\right)_{pS} \quad \text{......................................(9)}$$

It immediately follows from equation (9) that the lapse rate Γ in the ocean depends not only on the temperature but also on the changes in specific volume α of ocean waters with temperature. Thus, in order to calculate Γ, it remains to find out the expression for $\left(\dfrac{\partial \alpha}{\partial T}\right)_{pS}$ in calculating the lapse rate Γ; the rest of the analysis is devoted to this derivation. To begin with salinity and pressure remaining constant, the change of specific volume with respect to temperature is given by

$$\frac{\partial \alpha}{\partial T} = \frac{\partial}{\partial T}\left(\frac{1}{\rho}\right) = -\frac{1}{\rho^2}\frac{\partial \rho}{\partial T} = -\frac{1}{\rho}\left(\frac{1}{\rho}\frac{\partial \rho}{\partial T}\right)_{pS} \quad \dots\dots\dots\dots\dots\dots\dots\dots\dots\dots\dots\dots\dots(10)$$

From the equation of state in the differential form, $d\rho/\rho = \gamma_T\, dp - \alpha_T\, dT + \beta\, dS$, the thermal expression coefficient α_T has the following definition,

$$\alpha_T = -\frac{1}{\rho}\left(\frac{\partial \rho}{\partial T}\right)_{pS} \quad \dots\dots\dots\dots\dots\dots\dots\dots\dots\dots\dots\dots\dots\dots\dots(11)$$

Hence the expression (10) in the light of (11) becomes

$$\frac{\partial \alpha}{\partial T} = \frac{\alpha_T}{\rho} \quad \dots\dots\dots\dots\dots\dots\dots\dots\dots\dots\dots\dots\dots\dots\dots\dots(12)$$

Therefore the expression for the adiabatic lapse rate Γ can be written from (9) using (12) as

$$\Gamma = \frac{T\alpha_T}{\rho C_p} \quad \text{(T is Kelvin)} \quad \dots\dots\dots\dots\dots\dots\dots\dots\dots\dots\dots\dots\dots(13)$$

Equation (13) is the standard expression for the adiabatic lapse rate $\Gamma = \left(\dfrac{\partial T}{\partial p}\right)_{pS}$ in the ocean. The first determination of adiabatic lapse rate Γ, was done by Thompson (1857), who correctly showed the negative nature of Γ in cool fresh waters. The most modern measurements of Γ, made by Rogener and Soll [1980] and Caldwell and Eide (1980: Deep-sea Research, vol.27A), have confirmed that Γ is negative in cool fresh waters. A most recent account of Γ has been given by McDougall and Feistel (2003, Deep-Sea Research pt I, vol.50); it is a very informative paper which has been referred for preparing the material presented here.

Potential Temperature

The potential temperature with respect to a given reference pressure p_r is defined as

$$\theta = T + \int_{p}^{p_r} \Gamma(p', \theta[p, T, S; p'], S)\, dp' \quad \dots\dots\dots\dots\dots\dots\dots\dots\dots\dots\dots(14)$$

$\theta = \theta(p, T, S; p_r)$ is the potential temperature and $\Gamma = \Gamma(p, T, S)$ is the lapse rate. The recent technique of evaluating θ is to solve the equality

$$\eta(p_r, \theta, S) = \eta(p, T, S) \quad \dots\dots\dots\dots\dots\dots\dots\dots\dots\dots\dots\dots\dots\dots(15)$$

Substitute for Γ in (14) using the expression for Γ from eqn. (13); and we obtain $\theta = T + \int\limits_z^0 \dfrac{T}{\rho} \cdot \dfrac{\alpha_T}{C_p} \cdot g\rho dZ'$ and using the hydrostatic equilibrium , we can write

$$\theta = T + \int\limits_Z^0 \frac{T}{\rho} \cdot \frac{\alpha_T}{C_p} \cdot gpdZ'$$ ....(16)

The limits of integration are set accordingly; that is $Z' = Z$ at $p' = p$; and $Z' = 0$ at $p' = p$. By changing the sense of integration from 0 to Z, the eq. (16) can be written as

$$\theta = T - \int\limits_0^Z \frac{T.\alpha_T}{C_p} gdZ$$ ....(17)

The potential temperature θ of a layer of thickness δZ in the sea may be written as

$$\theta = T - \delta T \; ; \delta T = \int\limits_0^{\delta Z} \frac{T.\alpha_T}{C_p} gdZ' = \frac{T.\alpha_T}{C_p} g\delta Z$$ ....(18)

This derivation is due to Kelvin (1857). In the deep ocean, the potential temperature θ is less than the in situ temperature T by amount δT, which is mainly determined by the expansion coefficient α_T. Moreover, α_T can vary over a wider range and could even become negative for cool fresh water near freezing temperature. That is why lapse rate Γ is generally negative in the deep ocean. The correction δT obtained by Thomson (1857) is indeed a very remarkable finding for deep ocean waters (JJ Thompson later became Lord Kelvin). A notable point from the expression for Γ in (13) is that it varies linearly with α_T and T. Consequently if two parcels at same salinity but different temperatures are brought down adiabatically to deeper levels, the warm water parcel will warm up more than the cold water parcel. Such an increase in temperature is due to compression; as a consequence, the temperature of adiabatically moving parcels is not conserved.

This can be further illustrated with an example: Consider a water parcel moving downwards in the ocean to deeper layers from surface at a level where pressure is 500 bar (i.e. about to a depth of 5000 m). Thus $\delta p = 500m \times 10^5 Pa$; if the initial temperature of the water parcel is T = 288 K, then adiabatically moving parcel will have a change in temperature by δT given by

$$\delta T = \left(\frac{\partial T}{\partial p}\right)_{\eta S} \delta p = \Gamma \delta p = \frac{T\alpha_T}{\rho C_p}\delta p = \frac{T.\alpha_T}{C_p}g\delta Z = \frac{288 \times 2.2 \times 10^{-4}}{4200} \times 9.81 \times 5000 = 0.74K$$

with α_T = 2.2 x $10^{-4}K^{-1}$ and C_p = 4200 J $kg^{-1} K^{-1}$. It is very informative to know the magnitude of Γ can be obtained as

$$\frac{\delta T}{\delta p} = \frac{0.74(K)}{5000(db)} = 0.148K / (1000db)$$

It shows that essentially Γ is very different from actual vertical gradients of temperature in the ocean. *The principal reason for such a large difference is that large scale overturning in the ocean is not allowed; this means that water masses from deep levels at a location do not reach to the surface through direct vertical movements (so common in the atmosphere).* Instead, the water masses are formed in cold high latitudes, sink to deeper layer in the ocean along isopycnal surfaces associated with significant lateral movements. The outcropping of isopycnal surfaces in higher latitudes therefore plays a key role in the lateral movement of water masses from higher latitudes back to middle and low latitudes at deeper levels in the ocean. As a result of this lateral movement, large vertical gradients of temperature develop that are indeed observed in low and middle latitude ocean waters. This also explains the low salinity waters at deeper layers in the ocean. From this discussion it is evident that θ as a conserved quantity takes care of the adiabatic changes during vertical movements of water masses. It therefore assumes immense importance because it allows tagging of parcels during their movement.

Potential Density

The potential density of a fluid parcel at pressure P is the density that the parcel would acquire if adiabatically brought to a reference pressure P_0, often 1 bar (100 kPa). Whereas density changes with changing pressure, potential density of a fluid parcel is conserved as the pressure experienced by the parcel changes (provided no mixing with other parcels or net heat flux occurs). The concept is used in oceanography and (to a lesser extent) atmospheric science.

Potential density is a dynamically important property: for static stability potential density must decrease upward. If it doesn't, a fluid parcel displaced upward finds itself lighter than its neighbors, and continues to move upward; similarly, a fluid parcel displaced downward would be heavier than its neighbors. This is true even if the density of the fluid decreases upward. In stable conditions (potential density decreasing upward) motion along surfaces of constant potential density (isopycnals) is energetically favored over flow across these surfaces (diapycnal flow), so most of the motion within a 3-D geophysical fluid takes place along these 2-D surfaces.

In oceanography, the symbol ρ_θ is used to denote *potential density*, with the reference pressure P_0 taken to be the pressure at the ocean surface. The corresponding *potential density anomaly* is denoted by $\sigma_\theta = \rho_\theta - 1000\,\text{kg/m}^3$. Because the compressibility of seawater varies with salinity and temperature, the reference pressure must chosen to be near the actual pressure to keep the definition of potential density dynamically meaningful. Reference pressures are often chosen as a whole multiple of 100 bar; for water near a pressure of 400 bar (40 MPa), say, the reference pressure 400 bar would be used, and the potential density anomaly symbol would be written σ_4. Using these reference pressures can be defined the potential density surfaces, used in the analyses of ocean data and to construct models of the ocean current. Neutral density surfaces, defined using another variable called neutral density (γ^n), can be considered the continuous analog of these potential density surfaces.

Potential density adjusts for the effect of compression in two ways:

- The effect of a parcel's change in volume due to a change in pressure (as pressure increases, volume decreases).

- The effect of the parcel's change in temperature due to adiabatic change in pressure (as pressure increases, temperature increases).

A parcel's density may be calculated from an equation of state:

$$\rho = \rho(P,T,S_1,S_2,...)$$

where T is temperature, P is pressure and S_n are other tracers that affect density (e.g. salinity of seawater). The potential density would then be calculated as:

$$\rho_\theta = \rho(P_0,\theta,S_1,S_2,...)$$

where θ is the potential temperature of the fluid parcel for the same reference pressure P_0.

The adiabatic lapse rate Γ has been used to define the potential temperature θ in equation (14), it can also be used to define the potential density in the ocean. It has been averred earlier that the seawater is almost incompressible but in deeper levels the pressures are so high that density changes need to be taken in to account. And thus arises the concept of potential density which can be defined as the density a water parcel would attain if it were lifted adiabatically from a pressure level p where salinity is S, to a reference pressure p_r at Z = 0 (i.e. surface of the ocean). Thus, the potential density is defined as

$$\rho_\theta = \rho(p_r, \theta, S) \ (kg/m^3) \ ....(19)$$

The potential temperature is calculated using the expression given in (18).

Static Stability of Seawater

Under equilibrium conditions, the density ρ is a function of depth z alone and a column of seawater can be at rest. But, does this equilibrium stable? The stability of the ocean at rest can be tested by considering the exchange of two water parcels at different levels. If the displaced parcel finds itself heavier than it will sink back to its original level implying a stable ocean column. Both "adiabatic" and "isentropic" are the words which are used to refer to a process in which "heat content is not changed". Thus, "adiabatic" means "no exchange of heat with surroundings" ; where as "isentropic" means "no change of entropy". From the expression of adiabatic lapse rate Γ given in (13), the temperature change dT is related to the pressure change dp by the formula

$$dT = \Gamma dp = \frac{T\alpha_T}{\rho C_p} dp = \frac{\alpha T \alpha_T}{C_p} dp \ ....(20)$$

Using the hydrostatic equation (20) can be written as

$$dT = \frac{T\alpha_T}{\rho C_p}(-gpdZ) \Rightarrow dT = -\frac{gT\alpha_T}{C_p} dz = -\Gamma_a dz \ ....(21)$$

Where, Γ_a is the adiabatic lapse rate with depth in the ocean.

The parameter Γ_a can be expressed as

$$\Gamma_a = \frac{gT\alpha_T}{C_p} = \frac{T}{H_T}; H_T = \frac{C_P}{\alpha_T g} \ ....(22)$$

The changes in C_p due to salinity changes in the ocean are relatively small (~8% only). Further, for every 10 meter depth, pressure changes by 1 bar (10 decibar); hence pressures in the ocean column are generally expressed in decibar which coincide with ocean depths in meters. The changes in α_T (being in the denominator) has larger impact on Γ_a in deeper levels because α_T varies with depth. The temperature scale height $H_T \sim 2000$ km gives a value of Γ_a equal to 0.07 K km^{-1} when the ocean is at rest. Though H_T depends both on C_p and α_T the value of Γ_a largely depends upon α_T (being in the denominator) which varies with depth. It is worthwhile to remark here that for gases the product $T\alpha_T$ in (22) is equal to unity and relation becomes $\Gamma_a = g/C_p$.

Specific Heat C_p : The calculation of adiabatic lapse rate requires an accurate evaluation of C_p in eqn. (22). The specific heat at the surface of the ocean (where reference pressure is zero) is first calculated for freshwater as

$$C_p(0,T,0) = 4217.4 - 3.720283T = 0.1412855\,T^2 - 2.654387 \times 10^{-3}T^3 + 2.093236 \times 10^{-5}T^4 [J\ kg^{-1}\ K^{-1}] \quad ...(23)$$

The specific heat for seawater $C_p(0,T, S)$ at the ocean surface is then calculated as

$$C_p(0,T,S) = C_p(0,T,0) + S[-7.6444 + 0.107276\,T - 1.3839 \times 10^{-3}T^2] + S^{3/2}[0.17709 - 4.0772 \times 10^{-3}T + 5.3539 \times 10^{-5}T^2] \quad ..(24)$$

This formula can be checked against the result $C_p(0, 40, 40) = 3981.050$. The standard deviation is 0.074 for the algorithm fit (24). Finally the values of $C_p(p, T, S)$ are obtained by integrating the expression of $\left(\dfrac{\partial C_P}{\partial p}\right)_T$ which one can derive from the thermodynamic relations as shown below.

Expression for $\left(\dfrac{\partial C_P}{\partial p}\right)_T$:

For simplicity the seawater of uniform salinity is assumed so that in the equation of state of seawater. Neglecting the effect of evaporation (dn=0), the expression for internal energy change, de, from Gibbs relation (41) reads

de = Tdη - p dα or Tdη = de + p dα ....(25)

$$T\left(\frac{\partial \eta}{\partial T}\right)_p = \left(\frac{\partial e}{\partial T}\right)_p + p\left(\frac{\partial \alpha}{\partial T}\right)_p \qquad (Pressure\ cons\tan t) \qquad \qquad ....(26)$$

$$T\left(\frac{\partial \eta}{\partial p}\right)_T = \left(\frac{\partial e}{\partial p}\right)_T + p\left(\frac{\partial \alpha}{\partial p}\right)_T \qquad (Temperature\ cons\tan t) \qquad \qquad ....(27)$$

Differentiating (26) with respect to p, we get

$$T\left(\frac{\partial^2 \eta}{\partial p \partial T}\right)_p = \left(\frac{\partial^2 e}{\partial p \partial T}\right)_p + \left(\frac{\partial \alpha}{\partial T}\right)_p + p\left(\frac{\partial^2 \alpha}{\partial p \partial T}\right)_p \qquad ....(28)$$

Differentiating (27) with respect to T, we get

$$T\left(\frac{\partial^2 \eta}{\partial T \partial p}\right)_T + \left(\frac{\partial \eta}{\partial p}\right)_T = \left(\frac{\partial^2 e}{\partial T \partial p}\right)_T + p\left(\frac{\partial^2 \alpha}{\partial T \partial p}\right)_T \qquad ....(29)$$

At a point (T, p) on thermodynamic diagram, cross-derivatives of a thermodynamical variable are equal; this implies that

$$\left(\frac{\partial^2 \eta}{\partial T \partial p}\right)_T = \left(\frac{\partial^2 \eta}{\partial p \partial T}\right)_p \quad and \quad \left(\frac{\partial^2 e}{\partial p \partial T}\right)_p = \left(\frac{\partial^2 \alpha}{\partial T \partial p}\right)_T .$$

Subtracting (28) from (29) and using the equality of cross-derivatives at a point (T, p) on the thermodynamic diagram, one obtains the so-called Maxwell's relation,

$$\left(\frac{\partial \eta}{\partial p}\right)_T = -\left(\frac{\partial \alpha}{\partial T}\right)_p$$ ....(30)

Maxwell's equations can be derived easily from the thermodynamic equation given in (41) without considering the evaporation term; which reads

$$de = T \, d\eta - p \, d\alpha + \mu \, dS$$ ....(31)

From calculus, one writes the total differential as

$$de = \left(\frac{\partial e}{\partial \eta}\right)_{\alpha S} d\eta + \left(\frac{\partial e}{\partial \alpha}\right)_{\eta S} d\alpha + \left(\frac{\partial e}{\partial S}\right)_{\eta \alpha} dS$$ ....(32)

Comparing (31) and (32), one obtains

$$T = \left(\frac{\partial e}{\partial \eta}\right)_{\alpha S} ; p = -\left(\frac{\partial e}{\partial \alpha}\right)_{\eta S} ; \mu = \left(\frac{\partial e}{\partial S}\right)_{\eta \alpha}$$ ....(33)

Because the right-hand side of (31) is an exact differential, we have

$$\left(\frac{\partial^2 e}{\partial \eta \partial \alpha}\right) = \left(\frac{\partial^2 e}{\partial \alpha \partial \eta}\right)$$ ....(34)

Thus the first Maxwell Equation can be obtained from (33) and (34) as

$$-\left(\frac{\partial p}{\partial \eta}\right)_\alpha = \left(\frac{\partial T}{\partial \alpha}\right)_\eta$$ ....(35)

The specific heat for seawater C_p has been defined in eq. (8) which can be used to obtain the derivative of C_p with the pressure as follows,

$$C_p = T\left(\frac{\partial \eta}{\partial T}\right)_p \Rightarrow \left(\frac{\partial C_p}{\partial p}\right)_T = \left\{\frac{\partial}{\partial p}\left[T\left(\frac{\partial \eta}{\partial T}\right)_p\right]\right\}_T$$ ....(36)

This gives $\left(\dfrac{\partial C_p}{\partial p}\right)_T = \left\{T\dfrac{\partial}{\partial T}\left[\left(\dfrac{\partial \eta}{\partial p}\right)_T\right]\right\}_p$ on interchanging the orientation of the differentiation with

respect to p and T; and now substitute for $\left(\dfrac{\partial \eta}{\partial p}\right)_T$ from (30) we get,

$$\left(\frac{\partial C_p}{\partial p}\right)_T = -T\left(\frac{\partial^2 \alpha}{\partial T^2}\right)_p$$ ....(37)

Equation (37) permits us to obtain values of C_p at higher pressures while temperature remains constant, using the value of $C_p(0, T, S)$ that is calculated from relation (24). For the computation of α at higher pressures in the ocean, it is required to write Taylor expansion for $C_p(\delta p, T, S)$ as follows,

$$C_p(\delta p, T, S) = C_p(0, T, S) + \left(\frac{\partial C_p}{\partial p}\right)_T \delta p$$ ....(38)

The second term on the right hand side of (7.38) involves second derivatives of α, that is $\dfrac{\partial^2 \alpha}{\partial T^2}$; and it can be evaluated by putting δp for p in eq. (6.5), which gives

$$\alpha(\delta p, T, S) = \alpha(0, T, S)\left[1 - \frac{\delta p}{K_T(\delta p, T, S)}\right]$$ ....(39)

On obtaining $\left(\dfrac{\partial^2 \alpha}{\partial T^2}\right)$ from eqn. (7.39) keeping δp fixed and S constant, C_p at a depth (at pressure

δp) is obtained from eqn. (38). The value of $C_p(\delta p, T, S)$ gives the adiabatic lapse rate Γ_a. The calculation of second derivative of α is given in what follows (38).

Expression for α : Since $\alpha(\delta p, T, S)$ is a product of two functions, we can apply the differentiation of product of functions using the Leibniz rule in the following manner

$$\frac{\partial^2}{\partial T^2}\{f(T)g(T)\} = f''(T)g(T) + 2f'(T)g'(T) + f(T)g''(T)$$

and we obtain the following an expression for $\dfrac{\partial}{\partial}$

$$\frac{\partial^2 \alpha}{\partial T^2} = \frac{\partial^2 \alpha(0,T,S)}{\partial T^2}\left[1 - \frac{\delta p}{K_T(\delta p, T, S)}\right] + 2\frac{\partial \alpha(0,T,S)}{\partial T}\cdot\frac{\delta p}{K_T^2(\delta p, T, S)}\frac{\partial K_T(\delta p, T, S)}{\partial T}$$

$$+\alpha(0,T,S)\left[\frac{1}{K_T^2(\delta p, T, S)}\frac{\partial^2 K_T(\delta p, T, S)}{\partial T^2} - \frac{2}{K_T^2(\delta p, T, S)}\left(\frac{\partial K_T(\delta p, T, S)}{\partial T}\right)^2\right]$$ ....(40)

Further $\alpha(0, T, S)$ and $K_T(\delta p, T, S)$ are evaluated from the following expressions,

$$\frac{1}{\alpha(0,T,S)} = A + BS + CS^{3/2} + DS^2 = \frac{1}{\alpha_0}$$ ....(41)

$$K_T(\delta p, T, S) = E + F \times S + G \times S^{3/2} + (H + I \times S + J \times S^{3/2})\delta p + (M + N \times S)\delta p^2 \quad \text{................(42)}$$

From the expressions in (41) and (42) we can calculate $\dfrac{\partial^2 \alpha_0}{\partial T^2}, \dfrac{\partial \alpha_0}{\partial T}, \dfrac{\partial K_T}{\partial T}$ and $\dfrac{\partial^2 \alpha}{\partial T^2}$; On substituting their values in (40), one obtains the value of $\dfrac{\partial^2 \alpha}{\partial T^2}$, which allows determining from eqn. (37); and using this value of the derivative of C_p we obtain the value of p=δp at the level C_p. Finally the adiabatic lapse rate in the ocean is obtained at the level p=δp as

$$\Gamma_a = \frac{gT\,\alpha_T}{C_p(\delta p, T, S)} \quad \text{..(43)}$$

Gibbs Free Energy

In thermodynamics, the Gibbs free energy (IUPAC recommended name: Gibbs energy or Gibbs function; also known as free enthalpy to distinguish it from Helmholtz free energy) is a thermodynamic potential that can be used to calculate the maximum of reversible work that may be performed by a thermodynamic system at a constant temperature and pressure (isothermal, isobaric). Just as in mechanics, where the decrease in potential energy is defined as maximum useful work that can be performed, similarly different potentials have different meanings. The decrease in Gibbs free energy (J in SI units) is the *maximum* amount of non-expansion work that can be extracted from a thermodynamically closed system (one that can exchange heat and work with its surroundings, but not matter); this maximum can be attained only in a completely reversible process. When a system transforms reversibly from an initial state to a final state, the decrease in Gibbs free energy equals the work done by the system to its surroundings, minus the work of the pressure forces.

The Gibbs energy (also referred to as *G*) is also the thermodynamic potential that is minimized when a system reaches chemical equilibrium at constant pressure and temperature. Its derivative with respect to the reaction coordinate of the system vanishes at the equilibrium point. As such, a reduction in *G* is a necessary condition for the spontaneity of processes at constant pressure and temperature.

The Gibbs free energy, originally called *available energy*, was developed in the 1870s by the American scientist Josiah Willard Gibbs. In 1873, Gibbs described this "available energy" as

the greatest amount of mechanical work which can be obtained from a given quantity of a certain substance in a given initial state, without increasing its total volume or allowing heat to pass to or from external bodies, except such as at the close of the processes are left in their initial condition.

The initial state of the body, according to Gibbs, is supposed to be such that "the body can be made to pass from it to states of dissipated energy by reversible processes". In his 1876 magnum opus *On the Equilibrium of Heterogeneous Substances*, a graphical analysis of multi-phase chemical systems, he engaged his thoughts on chemical free energy in full.

Overview

According to the second law of thermodynamics, for systems reacting at STP (or any other fixed temperature and pressure), there is a general natural tendency to achieve a minimum of the Gibbs free energy.

The reaction $C_{(s)}^{diamond} \rightarrow C_{(s)}^{graphite}$ has a negative change in Gibbs free energy and is therefore thermodynamically favorable at 25°C and 1 atm. However, even though favorable, it is so slow that it is not observed. Whether a reaction is thermodynamically favorable does not determine its rate.

A quantitative measure of the favorability of a given reaction at constant temperature and pressure is the change ΔG (sometimes written "delta G" or "dG") in Gibbs free energy that is (or would be) caused by the reaction. As a necessary condition for the reaction to occur at constant temperature and pressure, ΔG must be smaller than the non-PV (e.g. electrical) work, which is often equal to zero (hence ΔG must be negative). ΔG equals the maximum amount of non-PV work that can be performed as a result of the chemical reaction for the case of reversible process. If the analysis indicated a positive ΔG for the reaction, then energy —in the form of electrical or other non-PV work— would have to be added to the reacting system for ΔG to be smaller than the non-PV work and make it possible for the reaction to occur.

The equation can be also seen from the perspective of the system taken together with its surroundings (the rest of the universe). First assume that the given reaction at constant temperature and pressure is the only one that is occurring. Then the entropy released or absorbed by the system equals the entropy that the environment must absorb or release, respectively. The reaction will only be allowed if the total entropy change of the universe is zero or positive. This is reflected in a negative ΔG, and the reaction is called exergonic.

If we couple reactions, then an otherwise endergonic chemical reaction (one with positive ΔG) can be made to happen. The input of heat into an inherently endergonic reaction, such as the elimination of cyclohexanol to cyclohexene, can be seen as coupling an unfavourable reaction (elimination) to a favourable one (burning of coal or other provision of heat) such that the total entropy change of the universe is greater than or equal to zero, making the *total* Gibbs free energy difference of the coupled reactions negative.

In traditional use, the term "free" was included in "Gibbs free energy" to mean "available in the form of useful work". The characterization becomes more precise if we add the qualification that it is the energy available for *non-volume* work. (An analogous, but slightly different, meaning of "free" applies in conjunction with the Helmholtz free energy, for systems at constant temperature).

However, an increasing number of books and journal articles do not include the attachment "free", referring to G as simply "Gibbs energy". This is the result of a 1988 IUPAC meeting to set unified terminologies for the international scientific community, in which the adjective 'free' was supposedly banished. This standard, however, has not yet been universally adopted.

History

The quantity called "free energy" is a more advanced and accurate replacement for the outdated term *affinity*, which was used by chemists in the earlier years of physical chemistry to describe the *force* that caused chemical reactions.

In 1873, Willard Gibbs published *A Method of Geometrical Representation of the Thermodynamic Properties of Substances by Means of Surfaces*, in which he sketched the principles of his new equation that was able to predict or estimate the tendencies of various natural processes to ensue when bodies or systems are brought into contact. By studying the interactions of homogeneous substances in contact, i.e., bodies composed of part solid, part liquid, and part vapor, and by using a three-dimensional volume-entropy-internal energy graph, Gibbs was able to determine three states of equilibrium, i.e., "necessarily stable", "neutral", and "unstable", and whether or not changes would ensue. Further, Gibbs stated:

> If we wish to express in a single equation the necessary and sufficient condition of thermodynamic equilibrium for a substance when surrounded by a medium of constant pressure p and temperature T, this equation may be written:
>
> $$\delta(\varepsilon - T\eta + pv) = 0$$
>
> when δ refers to the variation produced by any variations in the state of the parts of the body, and (when different parts of the body are in different states) in the proportion in which the body is divided between the different states. The condition of stable equilibrium is that the value of the expression in the parenthesis shall be a minimum.

In this description, as used by Gibbs, ε refers to the internal energy of the body, η refers to the entropy of the body, and v is the volume of the body.

Thereafter, in 1882, the German scientist Hermann von Helmholtz characterized the affinity as the largest quantity of work which can be gained when the reaction is carried out in a reversible manner, e.g., electrical work in a reversible cell. The maximum work is thus regarded as the diminution of the free, or available, energy of the system (*Gibbs free energy G* at T = constant, P = constant or *Helmholtz free energy F* at T = constant, V = constant), whilst the heat given out is usually a measure of the diminution of the total energy of the system (internal energy). Thus, G or F is the amount of energy "free" for work under the given conditions.

Until this point, the general view had been such that: "all chemical reactions drive the system to a state of equilibrium in which the affinities of the reactions vanish". Over the next 60 years, the term affinity came to be replaced with the term free energy. According to chemistry historian Henry Leicester, the influential 1923 textbook *Thermodynamics and the Free Energy of Chemical Substances* by Gilbert N. Lewis and Merle Randall led to the replacement of the term "affinity" by the term "free energy" in much of the English-speaking world.

Graphical Interpretation

Gibbs free energy was originally defined graphically. In 1873, American scientist Willard Gibbs published his first thermodynamics paper, "Graphical Methods in the Thermodynamics of Fluids", in which Gibbs used the two coordinates of the entropy and volume to represent the state of the body. In his second follow-up paper, "A Method of Geometrical Representation of the Thermodynamic Properties of Substances by Means of Surfaces", published later that year, Gibbs added in the third coordinate of the energy of the body, defined on three figures. In 1874, Scottish physicist James Clerk Maxwell used Gibbs' figures to make a 3D energy-entropy-volume thermodynamic surface of a fictitious water-like substance. Thus, in order to understand the very difficult concept of Gibbs free energy one must be able to understand its interpretation as Gibbs defined originally by section AB on his figure 3 and as Maxwell sculpted that section on his 3D surface figure.

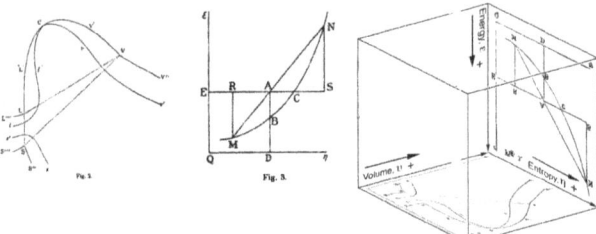

American scientist Willard Gibbs' 1873 figures two and three (above left and middle) used by Scottish physicist James Clerk Maxwell in 1874 to create a three-dimensional entropy (x), volume (y), energy (z) thermodynamic surface diagram for a fictitious water-like substance, transposed the two figures of Gibbs (above right) onto the volume-entropy coordinates (transposed to bottom of cube) and energy-entropy coordinates (flipped upside down and transposed to back of cube), respectively, of a three-dimensional Cartesian coordinates; the region AB being the first-ever three-dimensional representation of Gibbs free energy, or what Gibbs called "available energy"; the region AC being its capacity for entropy, what Gibbs defined as "the amount by which the entropy of the body can be increased without changing the energy of the body or increasing its volume.

Definitions

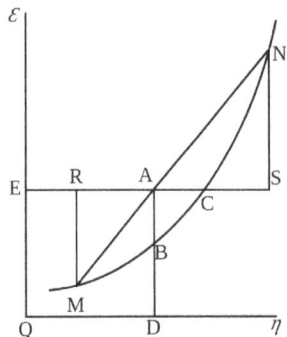

Willard Gibbs' 1873 available energy (free energy) graph, which shows a plane perpendicular to the axis of v (volume) and passing through point A, which represents the initial state of the body. MN is the section of the surface of dissipated energy. Qε and Qη are sections of the planes $\eta = 0$ and $\varepsilon = 0$, and therefore parallel to the axes of ε (internal energy) and η (entropy), respectively. AD and AE are the energy and entropy of the body in its initial state, AB and AC its *available energy* (Gibbs free energy) and its *capacity for entropy* (the amount by which the entropy of the body can be increased without changing the energy of the body or increasing its volume) respectively.

The Gibbs free energy is defined as:

$$G(p,T) = U + pV - TS$$

which is the same as:

$$G(p,T) = H - TS$$

where:

- U is the internal energy (SI unit: joule)

- p is pressure (SI unit: pascal)

- V is volume (SI unit: m³)

- T is the temperature (SI unit: kelvin)

- S is the entropy (SI unit: joule per kelvin)

- H is the enthalpy (SI unit: joule)

The expression for the infinitesimal reversible change in the Gibbs free energy as a function of its 'natural variables' p and T, for an open system, subjected to the operation of external forces (for instance electrical or magnetic) X_i, which cause the external parameters of the system a_i to change by an amount da_i, can be derived as follows from the First Law for reversible processes:

$$TdS = dU + pdV - \sum_{i=1}^{k} \mu_i dN_i + \sum_{i=1}^{n} X_i da_i + \cdots$$

$$d(TS) - SdT = dU + d(pV) - Vdp - \sum_{i=1}^{k} \mu_i dN_i + \sum_{i=1}^{n} X_i da_i + \cdots$$

$$d(U - TS + pV) = Vdp - SdT + \sum_{i=1}^{k} \mu_i dN_i - \sum_{i=1}^{n} X_i da_i + \cdots$$

$$dG = Vdp - SdT + \sum_{i=1}^{k} \mu_i dN_i - \sum_{i=1}^{n} X_i da_i + \cdots$$

where:

- μ_i is the chemical potential of the ith chemical component. (SI unit: joules per particle or joules per mole)

- N_i is the number of particles (or number of moles) composing the ith chemical component.

This is one form of Gibbs fundamental equation. In the infinitesimal expression, the term involving the chemical potential accounts for changes in Gibbs free energy resulting from an influx or outflux of particles. In other words, it holds for an open system. For a closed system, this term may be dropped.

Any number of extra terms may be added, depending on the particular system being considered. Aside from mechanical work, a system may, in addition, perform numerous other types of work. For example, in the infinitesimal expression, the contractile work energy associated with a thermodynamic system that is a contractile fiber that shortens by an amount $-dl$ under a force f would result in a term $f\,dl$ being added. If a quantity of charge $-de$ is acquired by a system at an electrical potential Ψ, the electrical work associated with this is $-\Psi de$, which would be included in the infinitesimal expression. Other work terms are added on per system requirements.

Each quantity in the equations above can be divided by the amount of substance, measured in moles, to form *molar Gibbs free energy*. The Gibbs free energy is one of the most important thermodynamic functions for the characterization of a system. It is a factor in determining outcomes such as the voltage of an electrochemical cell, and the equilibrium constant for a reversible reaction. In isothermal, isobaric systems, Gibbs free energy can be thought of as a "dynamic" quantity, in that it is a representative measure of the competing effects of the enthalpic and entropic driving forces involved in a thermodynamic process.

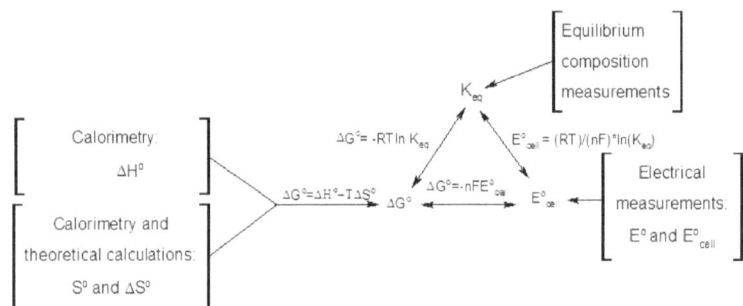

Relation to other relevant parameters

The temperature dependence of the Gibbs energy for an ideal gas is given by the Gibbs–Helmholtz equation and its pressure dependence is given by:

$$\frac{G}{N} = \frac{G^{\circ}}{N} + kT \ln \frac{p}{p^{\circ}}$$

if the volume is known rather than pressure then it becomes:

$$\frac{G}{N} = \frac{G^{\circ}}{N} + kT \ln \frac{V^{\circ}}{V}$$

or more conveniently as its chemical potential:

$$\frac{G}{N} = \mu = \mu^{\circ} + kT \ln \frac{p}{p^{\circ}}.$$

In non-ideal systems, fugacity comes into play.

Derivation

The Gibbs free energy total differential natural variables may be derived via Legendre transforms of the internal energy.

$$dU = TdS - pdV + \sum_i \mu_i dN_i.$$

The definition of G from above is

$$G = U + pV - TS.$$

Taking the total differential, we have

$$dG = dU + pdV + Vdp - TdS - SdT.$$

Replacing dU with the result from the first law gives

$$dG = TdS - pdV + \sum_i \mu_i dN_i + pdV + Vdp - TdS - SdT$$

$$= Vdp - SdT + \sum_i \mu_i dN_i$$

The natural variables of G are then p, T, and $\{N_i\}$.

Homogeneous Systems

Because S, V, and N_i are extensive variables, an Euler integral allows easy integration of dU:

$$U = TS - pV + \sum_i \mu_i N_i.$$

Because some of the natural variables of G are intensive, dG may not be integrated using Euler integrals as is the case with internal energy. However, simply substituting the above integrated result for U into the definition of G gives a standard expression for G:

$$G = U + pV - TS = (TS - pV + \sum_i \mu_i N_i) + pV - TS = \sum_i \mu_i N_i.$$

This result applies to homogeneous, macroscopic systems, but not to all thermodynamic systems.

Gibbs Free Energy of Reactions

To derive the Gibbs free energy equation for an isolated system, let S_{tot} be the total entropy of the isolated system, that is, a system that cannot exchange energy(heat and work) or mass with its surroundings. According to the second law of thermodynamics:

$$\Delta S_{tot} \geq 0$$

and if $\Delta S_{tot} = 0$ then the process is reversible. The heat transfer Q vanishes for an adiabatic system. Any adiabatic process that is also reversible is called an isentropic $\left(\dfrac{Q}{T} = \Delta S = 0\right)$ process.

Now consider a subsystem having internal entropy S_{int}. Such a system is thermally connected to its surroundings, which have entropy S_{ext}. The entropy form of the second law applies only to the closed system formed by both the system and its surroundings. Therefore, a process is possible only if

$$\Delta S_{int} + \Delta S_{ext} \geq 0.$$

If Q is the heat transferred to the system from the surroundings, then $-Q$ is the heat lost by the surroundings, so that $\Delta S_{ext} = -\dfrac{Q}{T}$, corresponds to the entropy change of the surroundings. We now have:

$$\Delta S_{int} - \frac{Q}{T} \geq 0$$

Multiplying both sides by T:

$$T\Delta S_{int} - Q \geq 0$$

Q is the heat transferred *to* the system; if the process is now assumed to be isobaric, then $Q = \Delta H$:

$$T\Delta S_{int} - \Delta H \geq 0$$

ΔH is the enthalpy change of reaction (for a chemical reaction at constant pressure). Then:

$$\Delta H - T\Delta S_{int} \leq 0$$

for a possible process. Let the change ΔG in Gibbs free energy be defined as

$\Delta G = \Delta H - T\Delta S_{int}$ *(eq.1)*

Notice that it is not defined in terms of any external state functions, such as ΔS_{ext} or ΔS_{tot}. Then the second law, which also tells us about the spontaneity of the reaction, becomes:

$\Delta G < 0$ favoured reaction (Spontaneous)

$\Delta G = 0$ Neither the forward nor the reverse reaction prevails (Equilibrium)

$\Delta G > 0$ disfavoured reaction (Nonspontaneous)

Gibbs free energy G itself is defined as

$G = H - TS_{int}$ *(eq.2)*

but notice that to obtain equation (1) from equation (2) we must assume that T is constant.

Thus, Gibbs free energy is most useful for thermochemical processes at constant temperature and pressure: both isothermal and isobaric. Such processes don't move on a P-T diagram, such as phase change of a pure substance, which takes place at the saturation pressure and temperature. Chemical reactions, however, do undergo changes in chemical potential, which is a state function. Thus, thermodynamic processes are not confined to the two dimensional P-V diagram. There is an additional dimension for the extent of the chemical reaction, associated with the changes of the amounts of the substances in the system. For the study of explosive chemicals, the processes are not necessarily isothermal and isobaric. For these studies, Helmholtz free energy is used.

If an isolated system ($Q = 0$) is at constant pressure ($Q = \Delta H$), then

$$\Delta H = 0$$

Therefore, the Gibbs free energy of an isolated system is

$$\Delta G = -T\Delta S$$

and if $\Delta G \leq 0$ then this implies that $\Delta S \geq 0$, back to where we started the derivation of ΔG.

Useful Identities to Derive the Nernst Equation

During a reversible electrochemical reaction at constant temperature and pressure, the following equations involving the Gibbs free energy hold:

$\Delta_r G = \Delta_r G^\circ + RT \ln Q_r$

$\Delta_r G^\circ = -RT \ln K$ (for a system at chemical equilibrium)

$\Delta_r G = w_{elec,rev} = -nFE$ (for a reversible electrochemical process at constant temperature and pressure)

$\Delta_r G^\circ = -nFE^\circ$ (definition of E°)

and rearranging gives

$$nFE^\circ = RT \ln K$$

$$nFE = nFE^\circ - RT \ln Q_r$$

$$E = E^\circ - \frac{RT}{nF} \ln Q_r$$

which relates the cell potential resulting from the reaction to the equilibrium constant and reaction quotient for that reaction (Nernst equation).

where

- $\Delta_r G$ = Gibbs free energy change per mole of reaction

- $\Delta_r G°$ = Gibbs free energy change per mole of reaction for unmixed reactants and products at standard conditions

- R = gas constant

- T = absolute temperature (in K)

- ln = natural logarithm

- Q_r = reaction quotient (unitless)

- K = equilibrium constant (unitless)

- $w_{elec,rev}$ = electrical work in a reversible process (chemistry sign convention)

- n = number of moles of electrons transferred in the reaction

- F = Faraday constant = 96485 C/mol (charge per mole of electrons)

- E = cell potential (in V)

- $E°$ = standard cell potential (in V)

Moreover, we also have:

$$K_{eq} = e^{-\frac{\Delta_r G°}{RT}}$$

$$\Delta_r G° = -RT(\ln K_{eq}) = -2.303 RT(\log_{10} K_{eq})$$

which relates the equilibrium constant with Gibbs free energy.

The Second Law of Thermodynamics and Metabolism

A chemical reaction will (or can) proceed spontaneously if the change in the total entropy of the universe that would be caused by the reaction is nonnegative. As discussed in the overview, if the temperature and pressure are held constant, the Gibbs free energy is a (negative) proxy for the change in total entropy of the universe. It is "negative" because S appears with a negative coefficient in the expression for G, so the Gibbs free energy moves in the opposite direction from the total entropy. Thus, a reaction with a positive Gibbs free energy will not proceed spontaneously. However, in biological systems (among others), energy inputs from other energy sources (including the sun and exothermic chemical reactions) are "coupled" with reactions that are not entropically favored (i.e. have a Gibbs free energy above zero). Taking into account the coupled reactions, the total entropy in the universe increases. This coupling allows endergonic reactions, such as photosynthesis and DNA synthesis, to proceed without decreasing the total entropy of the universe. Thus biological systems do not violate the second law of thermodynamics.

Standard Energy Change of Formation

The standard Gibbs free energy of formation of a compound is the change of Gibbs free energy that accompanies the formation of 1 mole of that substance from its component elements, at their standard states (the most stable form of the element at 25 degrees Celsius and 100 kilopascals). Its symbol is $\Delta_f G°$.

All *elements* in their standard states (diatomic oxygen gas, graphite, etc.) have standard Gibbs free energy change of formation equal to zero, as there is no change involved.

$\Delta_f G = \Delta_f G° + RT \ln Q_f$; Q_f is the reaction quotient.

At equilibrium, $\Delta_f G = 0$ and $Q_f = K$ so the equation becomes $\Delta_f G° = -RT \ln K$; K is the equilibrium constant.

Table of Selected Substances

Substance	State	$\Delta_f G°$(kJ/mol)	$\Delta_f G°$(kcal/mol)
NO	g	87.6	20.9
NO_2	g	51.3	12.3
N_2O	g	103.7	24.78
H_2O	g	-228.6	-54.64
H_2O	l	-237.1	-56.67
CO_2	g	-394.4	-94.26
CO	g	-137.2	-32.79
CH_4	g	-50.5	-12.1
C_2H_6	g	-32.0	-7.65
C_3H_8	g	-23.4	-5.59
C_6H_6	g	129.7	29.76
C_6H_6	l	124.5	31.00

Gibbs Function

The thermodynamic relation (31), $de = T\, d\eta - p\, d\alpha + \mu\, dS$, maybe written as

$$de = d(T\eta) - d(p\alpha) - \eta dT + \alpha dp + \mu dS \dots\dots(44)$$

On rearranging this equation, we get the following relation

$$d(e - T\eta + p\alpha) = \eta dT + \alpha dp + \mu dS \dots\dots(45)$$

Define the Gibbs potential (also called the Gibbs function and Gibbs free energy) as

$$G = e - T\eta + p\alpha \dots\dots(46)$$

The eqn. (45) takes the form,

$$dG = \eta dT + \alpha dp + \mu dS \dots\dots(47)$$

Since we formally have,

$$dG = \left(\frac{\partial G}{\partial T}\right) dT + \left(\frac{\partial G}{\partial p}\right) dp + \left(\frac{\partial G}{\partial S}\right) dS$$

$$\dots\dots(48)$$

We may write from a comparison of (47) and (48) the following definitions,

$$\left(\frac{\partial G}{\partial T}\right)_{P,S} = -\eta; \quad \left(\frac{\partial G}{\partial p}\right)_{T,S} = \alpha \ \text{ and } \ \left(\frac{\partial G}{\partial S}\right)_{T,\eta} = \mu$$

....(49)

Oceanographers have obtained a highly accurate equation of state of seawater and relations for evaluating other physical properties of seawater. The form of the Gibbs function that accurately reproduces the properties of seawater with certain degree of economy has been given by Vallis (2006). Though the expression is rather complicated yet it offers sufficient ease in computations; the expression for the Gibbs function is,

$$G = G_0 - \eta_0(T - T_0) + \mu(S - S_0) - C_{p0}T\left[\ln\left(\frac{T}{T_0}\right) - 1\right][1 + \beta_s^*(S - S_0)]$$

$$+\alpha_0(p - p_0)\left[1 + \beta_T(T - T_0) - \beta_s(S - S_0) - \frac{\beta_p}{2}(p - p_0)\right]$$

$$+\alpha_0(p - p_0)\left[1 + \beta_T\frac{\beta_T\gamma^*}{2}(p - p_0)(T - T_0) + \frac{\beta_T^*}{2}(T - T_0)^2\right]$$

...........................(50)

The specific volume $\alpha = \left(\dfrac{\partial G}{\partial p}\right)_{T,S}$ is given by the following expression,

$$\alpha = \alpha_0[1 + \beta_T(1 + \gamma^* p)(T - T_0) + \frac{\beta_T^*}{2}(T - T_0)^2 - \beta_S(S - S_0) - \beta_p p]$$

...............................(51)

The parameters appearing in (63) have the following meaning and values :

ρ_0	Reference density	1.027 x 10³ kg/m³
α_0	Reference specific volume	9.738 E-4 m³ /kg
T_0	Reference temperature	283 K
S_0	Reference Salinity	35 psu ≈35%
C_{s0}	Reference Sound Speed	1490 m/s
β_T	First Thermal Expansion Coefficient	1.67 E-4 K⁻¹
β_T^*	Second Thermal Expansion Coefficient	1.00 E-5 K⁻²
β_s	Haline contraction Coefficient	0.78 E-3 psu⁻¹
β_p	Compressibility coefficient (=α_0 / C_{s0}^2)	4.39 E-10 ms²/kg
γ^*	Thermobaric parameter	1.1 E-8 Pa⁻¹
C_{po}	Specific heat capacity at constant pressure	3986 J/kg/K
β_s^*	Haline heat capacity Coefficient	1.5 E-3 psu⁻¹
$P_0=0$	Pressure at sea surface	0.0 Pa

The expression of entropy $\eta = -\left(\dfrac{\partial G}{\partial T}\right)_{P,S}$ is given by

$$\eta = \eta_0 + C_{po}\ln\left(\frac{T}{T_0}\right)[1 + \beta_s^*(S - S_0)] - \alpha_0 p\left[\beta_T + \beta_T\gamma^*\frac{p}{2} + \beta_T^*(T - T_0)\right]$$

....................(52)

The specific heat capacity, $c_p = T\left(\dfrac{\partial \eta}{\partial T}\right)_p$ is given by

$$C_p = C_{po}\left[1 + \beta_s^*(S - S_0)\right] - \alpha_o p \beta_t^* T \qquad \dots\dots\dots\dots\dots\dots\dots\dots\dots\dots\dots\dots\dots\dots\dots\dots(53)$$

To a first approximation C_p is a constant that varies mildly with salinity and also depends weakly on temperature and pressure.

The thermal expansion coefficient, $a_T = \dfrac{1}{\alpha}\left(\dfrac{\partial \alpha}{\partial T}\right)_{p,S}$ is given by

$$\alpha_T = \left(\frac{\alpha_0}{\alpha}\right)\left[\beta_T + \beta_T \gamma^* p + \beta_T^*(T - T_0)\right] \qquad \dots\dots\dots\dots\dots\dots\dots\dots\dots\dots\dots\dots\dots(54)$$

where α, the specific volume, has been calculated using the expression (51). Finally, the adiabatic lapse rate Γ is obtained by substituting $\left(\dfrac{\partial \alpha}{\partial T}\right)_{p,S}$ from (54) as

$$\Gamma_a = \left(\frac{\partial T}{\partial p}\right)_\eta = \frac{T}{C_p}\left(\frac{\partial \alpha}{\partial T}\right)_{pS} = \frac{T}{C_p}\alpha_0\left[\beta_T\left(1 + \gamma^* p\right) + \beta_T^*(T - T_0)\right] \qquad \dots\dots\dots\dots\dots\dots\dots\dots\dots(55)$$

This is a simpler expression to obtain Γ_a but accuracy of the lapse rate nevertheless depends on the accuracy of C_p. Note that all the expressions derived here are approximate. Though there are some errors involved, yet their magnitude are small enough, so the final values can be confidently used in calculations.

For obtaining an expression for adiabatic lapse rate we can also use the following expression for entropy change $d\eta$ which is largely determined by temperature and pressure changes in the ocean,

$$d\eta = \left(\frac{\partial \eta}{\partial T}\right)_{pS} dT + \left(\frac{\partial \eta}{\partial p}\right)_{TS} dp$$

On setting $d\eta = 0$ in the above equation, we get the following relation for the lapse rate,

$$\left(\frac{\partial T}{\partial P}\right) = -\left(\frac{\partial \eta}{\partial p}\right)_{TS}\left(\frac{\partial \eta}{\partial T}\right)_{pS}^{-1} ; \left(\frac{\partial \eta}{\partial T}\right)_{pS} = -\frac{C_p}{T} ; \left(\frac{\partial \eta}{\partial p}\right)_{TS} = -\left(\frac{\partial \alpha}{\partial T}\right)$$

The above relations give

$$\left(\frac{\partial T}{\partial P}\right) = \frac{T}{C_p} \times \left(\frac{\partial \alpha}{\partial T}\right) = \frac{T\alpha}{C_p} \times \left(\frac{1}{\alpha}\frac{\partial \alpha}{\partial T}\right)$$

Hence the expression of Γ_a is obtained as

$$\Gamma_a = \left(\frac{\partial T}{\partial p}\right) = \frac{T\alpha_T}{\rho C_p}$$

This expression for Γ_a is same as (13) but obtained in much simpler manner.

Double-diffusion

The ocean columns are stably stratified because density increases with depth and the changes in density of seawater are mainly determined by the changes in temperature and salinity. It has also been emphasized that different water masses, be of same density but having different temperatures and salinities, produce denser waters after mixing under the mechanical forcing of winds at the surface. However, a counterintuitive physical phenomenon was observed by M.E. Stern (The "salt fountain" and thermohaline convection, Tellus, vol.12, 1960), which arises due to the differences in the molecular diffusivities of salt and temperature. This is known as the "double-diffusion" phenomenon. Why the mixing caused by double-diffusion is counterintuitive? Because, due this kind of mixing, dense waters become denser and light waters become lighter. If one considers a warm and salty water layer above a relatively colder and less salty water layer in the ocean, then due to the phenomenon of double-diffusion, the interface becomes unstable even though the density of the upper layer is lesser than that of the lower. The main reason being that molecular diffusion of heat is 100 times more than that of the salt, and the interfacial instability is caused by this difference in molecular diffusivities of the two important components of seawater viz., heat and salt. Ocean circulation is dominated both by large scale motions and eddies. Therefore, at this point, two important questions arise:

- Does the large-scale dynamics of ocean circulation affect double-diffusion?

- Does double-diffusion affect the large-scale ocean circulation and thereby produce consequences of interest to the sensitivity of climate system?

Answer to the first question is relatively clear because the instability associated with double-diffusion is mostly confined to the interface, hence the scales of inhomogeneity that dominate the large-scale flow would hardly affect the dynamics operating at the double-diffusion. However, an answer to the second question has been investigated in recent years theoretically, numerically and also observationally. New observations from "Salt Finger Tracer Release Experiment" have thrown new light on the mixing induced by double-diffusion in transformation of water masses. When water mass transformation happens due to salt fingering, it impacts the climate system. The early view that double-diffusion has an important role in maintaining the thermocline, also favours the argument that double-diffusion has tangible influence on large-scale phenomena in the ocean. Finally, it may be mentioned that the adiabatic mechanisms that control the stratification and limit overturning of water masses would inevitably reduce the mixing induced by turbulence or salt fingering. This has been observed for the Antarctic Circumpolar Current (ACC), where adiabatic effects control stratification and overturning, because any land barrier does not obstructed it; consequently the flow reenters into the sameregion after completing the latitude circle.

Potential Temperature Definition in Numerical Models

The expression derived for various terms of thermodynamical variables that appear in the equation of state, allow calculations of lapse rate, enthalpy and potential temperature that are computationally complex. The efficiency of computations can however be achieved with a simplified equation of state. The ocean numerical models also use potential temperature as one of the thermodynamic variables, which can be obtained at a given instant solving the equation (15), i.e.

$$\eta(p_r, \theta, S) = \eta(p, T, S)$$

In the above equation $\eta(p, \theta, S)$, is obtained from (52) by putting $p = p_r$ we have

$$\eta(p_r, \theta, S) = \eta_o + C_{po} \ln\left(\frac{\theta}{T_0}\right)\left[1 + \beta_s^*(S - S_0)\right] - \alpha_o p_r\left[\beta_T + \beta_T \gamma^* \frac{P_r}{2} + \beta_T^*(\theta - T_o)\right]$$

(56) By taking $p_r = 0$, the above equation becomes

$$\eta(0, \theta, S) = \eta_o + C_{Po} \ln\left(\frac{\theta}{T_0}\right)\left[1 + \beta_s^*(S - S_0)\right]$$ ....(57)

Equation (15) implies that (57) must be equated to (52), which gives

$$C_{po} \ln\left(\frac{\theta}{T_0}\right)\left[1 + \beta_s^*(S - S_0)\right] = C_{Po} \ln\left(\frac{T}{T_0}\right)\left[1 + \beta_s^*(S - S_0)\right] - \alpha_o p\left[\left[\beta_T + \beta_T \gamma^* \frac{p}{2} + \beta_T^*(T - T_o)\right]\right]$$

The above expression can be put in the following concise form,

$$C_{po}\left[1 + \beta_s^*(S - S_0)\right]\ln\left(\frac{\theta}{T}\right) = -\alpha_o \beta_T p\left[1 + \frac{\gamma^*}{2}p + \frac{\beta_T^*}{\beta_t}(T - T_0)\right]$$ ....(58)

From the above expression, we can write the most amenable form of the potential temperature for numerical computations as,

$$\theta = T \exp\left\{-\frac{\alpha_o \beta_T p}{C_{po}\left[1 + \beta_s^*(S - S_0)\right]}\left[1 + \frac{\gamma^*}{2}p + \frac{\beta_T^*}{\beta_t}(T - T_0)\right]\right\}$$ ....(59)

The equation (59) is much like the definition of θ in the atmosphere thermodynamics. In writing (59), the deviations of T and θ from T_o are presumed to be small and that the exponential term itself is small. The quantity $C_p' = C_{Po}\left[1 + \beta_s^*(S - S_0)\right]$ is nearly constant; therefore a simplified form (59) for θ could be derived. If we write T' = T - T_o; θ' = θ - T_0 with θ_o = T_0 at the sea surface, then from eq. (59) we have

$$T = \theta\left\{1 + \frac{\alpha_o \beta_T p}{C_p'}\left(1 + \frac{\gamma^* P}{2}\right) + \frac{\alpha_o \beta_T^* p}{C_p'}T' + \frac{\alpha_o^2 \beta_T^{*2} p^2}{C_p'^2}T' + \frac{\alpha_o^2 \beta_T^{*2} p^2}{C_p'^2}(1 + \gamma^* p)^2 + ...\right\}$$

Substitute T = T_0 + T' ; θ = T_0 + θ' then we obtain a relation between the perturbations of T and θ as

$$T'\left(1 - \frac{\alpha_o \beta_T^* p}{C_{po}}\right) \approx \frac{\alpha_o T_o \beta_T}{C_{po}}p\left(1 + \frac{\gamma^* p}{2}\right) + \theta'\left(1 + \frac{\alpha_o \beta_T^*}{C_{po}}p\right)$$ ....(60)

The above expression can be approximated as

$$T' = \frac{\alpha_o T_o \beta_T}{C_{po}}p + \theta' \quad or \quad \theta' = T' - \frac{\alpha_o T_o \beta_T}{C_{po}}p$$ ....(61)

The coefficient of p in (61) is same as the lapse rate Γ, defined earlier in this section. The numerical models use potential temperature as a thermodynamic variable, though the potential enthalpy should be preferable choice like in the compressible models of atmosphere in motion. The main advantage of using potential enthalpy as a variable is that its interpretation is a useful measure of heat content and it is also a conserved quantity. For numerical models, the simplified form of the equation of state is given as

$$\alpha = \alpha_o \left[1 + \frac{\alpha_o p}{C_{So}'^2} + \beta_T \left(1 + \tilde{\gamma}^* p \right) \theta' + \frac{1}{2} \beta_T^* \theta'^2 - \beta_s \left(S - S_o \right) \right] \quad \dots\dots\dots\dots\dots(62)$$

$$\tilde{\gamma}^* = \gamma^* + \frac{\alpha_o \beta_T^* T_o}{C_{po}} \approx \gamma^* \, , \tilde{C}_{so}^{-2} = C_{so}^{-2} - \frac{\beta_T^2 T_o}{C_p} \approx C_{so}^{-2}$$

The equation (62) can be approximated further with the use of hydrostatic pressure $p = -\frac{g(z - z_0)}{\alpha_0}$; at $z = z_0$, $p = 0$ Thus one may write (62) as

$$\alpha = \alpha_0 \left[1 + \frac{g(z - z_o)}{C_{so}^2} + \beta_T \left(1 - \tilde{\gamma}^* \frac{g(z - z_o)}{\alpha_0} \right) \theta' + \frac{1}{2} \beta_T^* \theta'^2 - \beta_s (S - S_o) \right] \quad \dots\dots\dots(63)$$

Thus, we may use either (62) or (63) as the equation of the state, which simplified but it involves quadratic terms in the perturbation of potential temperature.

A final word about inclusion of various mixing mechanisms in numerical models should be in order. The mixing-driven circulation is certainly a correction to adiabatically controlled modes of circulation; therefore, numerical ocean models include various mixing mechanisms for simulating realistically the ocean currents. Double-diffusion has also been included in some models mainly to isolate microstructures developing due to its mixing effects in large-scale flows that are dynamically distinct from those arising due to turbulent mixing. In this manner, the relative roles of these processes could be identified in the large scale dynamics of ocean currents.

Ocean Circulation and Surface Processes

Winds are produced in response to radiative heating of the atmosphere. These winds constitute an important forcing for ocean currents, which are generated due to momentum transfer into the ocean by winds. The pressure gradients generated by radiative heating could produce wind speeds of about 10 ms^{-1} in the atmosphere just above the ocean. Yet, there will be no momentum transfer by winds to ocean layers if there were no friction at the surface. Because of the frictional contact, *no slip condition* will be satisfied by the airflow at solid surface boundary; that is, the air in immediate contact with the boundary attains zero velocity. This will set up a velocity gradient (or shear) near the solid boundary. The shear flow set up in this manner is not stable at higher wind speeds because small disturbances can grow at the expense of mean motion to turn the flow turbulent. The turbulent eddies are responsible for the gusty nature of the flow, modify shear for a well-defined mean velocity structure to develop after sufficiently long time. The flow velocity is a function of z, i.e. the distance from the surface in vertical direction. The shear depends on mean stress τ, density ρ and distance z from the ground, and one obtains a logarithmic mean velocity profile.

Near the ground, wind shear varies as $(\partial u/\partial z) \sim 1/z$; that is, the inverse law holds only sufficiently close to the ground. The turbulent eddies in the shear flow close to the surface affect the transfer of momentum, heat and moisture at the interface of ocean and atmosphere. What are the other processes that affect the evolution of the surface layer besides the dynamical forcing of winds? How are their effects included in the analysis? There exists an extensive literature for an elaborate response to these questions. The other thermodynamically mediated processes are primarily the solar heating deposited in a few tens of meters in the upper ocean; evaporation at the sea surface; cooling by evaporation and sensible heat transfer at the surface of the ocean. Atmospheric circulations of different temporal and spatial scales produce horizontal gradients in the seawater properties in the open ocean affecting the current strengths below the interface. Such horizontal gradients, in conformity with nonlinearity of the equation of state of seawater,could impact the ocean stratification in different latitudes.

The ocean waters are inherently stratified with higher density waters arranged at the ocean bottom where temperatures are cold and salinity is low. The uniform distribution of potential temperatures in deep ocean signifies constant mixing forced by friction in the bottom Ekman layer. At the ocean surface, atmospheric processes constantly change the distribution of sea surface temperature and salinity through horizontal transport and stirring by winds, rain and evaporation, while moving air parcels pick up moisture from the sea surface and distribute it around the globe.

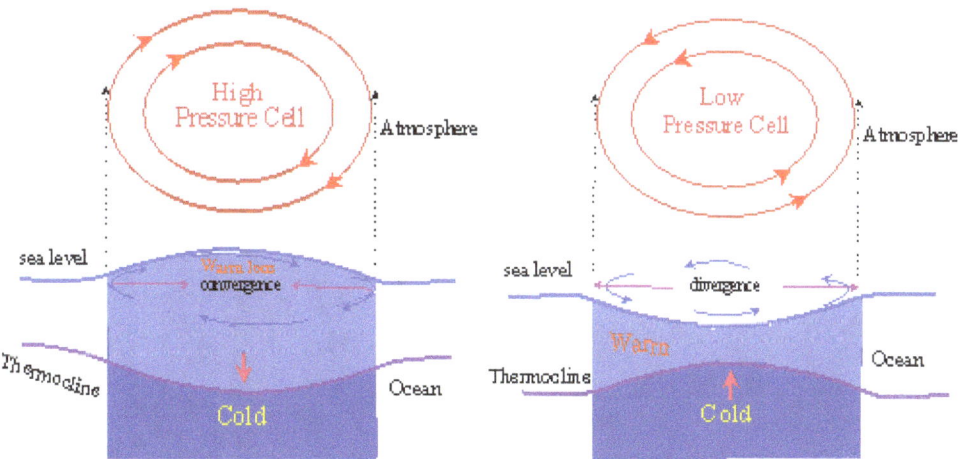

Development of a warm lens on the sea surface under the action of anticyclonic wind stress and cooling of surface waters under cyclonic wind stress. Note the vertical movement of thermocline under the influence of a circulation associated with a low and a high in the atmosphere.

For the same solar heating rate, land temperatures rise faster in summer than the sea temperatures and also land cools rapidly during winter with the diminishing radiative heating. As a result of this contrast, low-pressure systems during summers and high pressures during winters develop over land areas. Such a flip-flop in atmospheric pressures over land will result in the alteration of pressure systems over the oceans as well. For example, there is a high-pressure system (the Azores high) in the north Atlantic in summer, which is replaced by the Icelandic low during winter. Since the winds will be almost parallel to isobars over the oceanic regions, the wind induced surface ocean currents will have anticyclonic orientation below the atmospheric high pressure cell and cyclonic below the low pressure cell. Due to Ekman transport, light waters are on the right relative to the direction of the wind in the northern hemisphere, a lens of warm water will form the core of anticyclonic surface current cell, and a cold core for the cyclonic cell. In this manner, interaction forced by atmospheric circulation at the ocean surface contributes to variability of the climate system, which will finally affect the transfer processes at the ocean surface wherever air interacts with the sea.

These processes also play a dominant role in changing the density of seawater, and any loss of buoyancy would engender convective overturning in the ocean depth. From the equation of the state of seawater, it may be inferred that temperature and salinity generally play an opposing role on density of ocean waters. In the context of dynamics of oceans and atmosphere, it may be said that atmospheric winds drive ocean circulation; and evaporation from sea surface contributes water vapour to the atmosphere, which drives atmospheric circulation in combination with the radiative heating and convective overturning in the troposphere. Therefore, interaction of the oceans and atmosphere is key to the understanding of climate and its variability at shorter time scales and to climate change at longer time scales.

Drag Coefficient

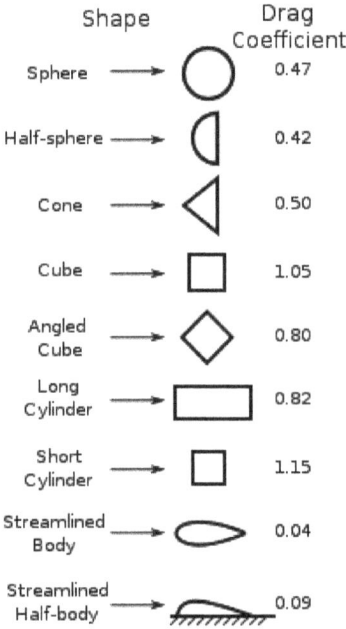

Measured Drag Coefficients

Drag coefficients in fluids with Reynolds number approximately 10^4

In fluid dynamics, the drag coefficient (commonly denoted as: C_d, C_x or C_w) is a dimensionless quantity that is used to quantify the drag or resistance of an object in a fluid environment, such as air or water. It is used in the drag equation in which a lower drag coefficient indicates the object will have less aerodynamic or hydrodynamic drag. The drag coefficient is always associated with a particular surface area.

The drag coefficient of any object comprises the effects of the two basic contributors to fluid dynamic drag: skin friction and form drag. The drag coefficient of a lifting airfoil or hydrofoil also includes the effects of lift-induced drag. The drag coefficient of a complete structure such as an aircraft also includes the effects of interference drag.

Definition

The drag coefficient c_d is defined as

$$c_d = \frac{2F_d}{\rho u^2 A}$$

where:

F_d is the drag force, which is by definition the force component in the direction of the flow velocity,

ρ is the mass density of the fluid,

u is the flow speed of the object relative to the fluid,

A is the reference area.

The reference area depends on what type of drag coefficient is being measured. For automobiles and many other objects, the reference area is the projected frontal area of the vehicle. This may not necessarily be the cross sectional area of the vehicle, depending on where the cross section is taken. For example, for a sphere $A = \pi r^2$ (note this is not the surface area = $4\pi r^2$).

For airfoils, the reference area is the nominal wing area. Since this tends to be large compared to the frontal area, the resulting drag coefficients tend to be low, much lower than for a car with the same drag, frontal area, and speed.

Airships and some bodies of revolution use the volumetric drag coefficient, in which the reference area is the square of the cube root of the airship volume (volume to the two-thirds power). Submerged streamlined bodies use the wetted surface area.

Two objects having the same reference area moving at the same speed through a fluid will experience a drag force proportional to their respective drag coefficients. Coefficients for unstreamlined objects can be 1 or more, for streamlined objects much less.

Background

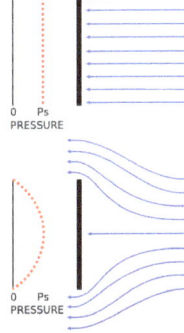

Flow around a plate, showing stagnation.

The drag equation

$$F_d = \tfrac{1}{2}\rho u^2 c_d A$$

is essentially a statement that the drag force on any object is proportional to the density of the fluid and proportional to the square of the relative flow speed between the object and the fluid.

C_d is not a constant but varies as a function of flow speed, flow direction, object position, object size, fluid density and fluid viscosity. Speed, kinematic viscosity and a characteristic length scale of the object are incorporated into a dimensionless quantity called the Reynolds number Re. C_d is thus a function of Re. In a compressible flow, the speed of sound is relevant, and C_d is also a function of Mach number Ma.

For certain body shapes, the drag coefficient C_d only depends on the Reynolds number Re, Mach number Ma and the direction of the flow. For low Mach number Ma, the drag coefficient is independent of Mach number. Also, the variation with Reynolds number Re within a practical range of interest is usually small, while for cars at highway speed and aircraft at cruising speed, the incoming flow direction is also more-or-less the same. Therefore, the drag coefficient C_d can often be treated as a constant.

For a streamlined body to achieve a low drag coefficient, the boundary layer around the body must remain attached to the surface of the body for as long as possible, causing the wake to be narrow. A high *form drag* results in a broad wake. The boundary layer will transition from laminar to turbulent if Reynolds number of the flow around the body is sufficiently great. Larger velocities, larger objects, and lower viscosities contribute to larger Reynolds numbers.

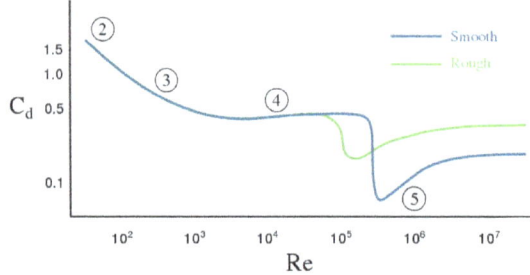

Drag coefficient C_d for a sphere as a function of Reynolds number Re, as obtained from laboratory experiments. The dark line is for a sphere with a smooth surface, while the lighter line is for the case of a rough surface. The numbers along the line indicate several flow regimes and associated changes in the drag coefficient:

•2: attached flow (Stokes flow) and steady separated flow,

•3: separated unsteady flow, having a laminar flow boundary layer upstream of the separation, and producing a vortex street,

•4: separated unsteady flow with a laminar boundary layer at the upstream side, before flow separation, with downstream of the sphere a chaotic turbulent wake,

•5: post-critical separated flow, with a turbulent boundary layer.

For other objects, such as small particles, one can no longer consider that the drag coefficient C_d is constant, but certainly is a function of Reynolds number. At a low Reynolds number, the flow around the object does not transition to turbulent but remains laminar, even up to the point at which it separates from the surface of the object. At very low Reynolds numbers, without flow separation, the drag force F_d is proportional to v instead of v^2 ; for a sphere this is known as Stokes law. The Reynolds number will be low for small objects, low velocities, and high viscosity fluids.

A C_d equal to 1 would be obtained in a case where all of the fluid approaching the object is brought to rest, building up stagnation pressure over the whole front surface. The top figure shows a flat plate with the fluid coming from the right and stopping at the plate. The graph to the left of it shows equal pressure across the surface. In a real flat plate, the fluid must turn around the sides, and full stagnation pressure is found only at the center, dropping off toward the edges as in the lower figure and graph. Only considering the front side, the C_d of a real flat plate would be less than 1; except that there will be suction on the back side: a negative pressure (relative to ambient). The overall C_d of a real square flat plate perpendicular to the flow is often given as 1.17. Flow patterns and therefore C_d for some shapes can change with the Reynolds number and the roughness of the surfaces.

Drag Coefficient c_d Examples

General

In general, c_d is not an absolute constant for a given body shape. It varies with the speed of airflow (or more generally with Reynolds number Re). A smooth sphere, for example, has a c_d that varies from high values for laminar flow to 0.47 for turbulent flow. Although the drag coefficient decreases with increasing Re , the drag force increases.

Aircraft

As noted above, aircraft use their wing area as the reference area when computing c_d , while automobiles (and many other objects) use frontal cross sectional area; thus, coefficients are not directly comparable between these classes of vehicles. In the aerospace industry, the drag coefficient is sometimes expressed in drag counts where 1 drag count = 0.0001 of a C_d .

Aircraft	
c_d	Aircraft type
0.021	F-4 Phantom II (subsonic)
0.022	Learjet 24
0.024	Boeing 787
0.0265	Airbus A380
0.027	Cessna 172/182
0.027	Cessna 310

0.031	Boeing 747
0.044	F-4 Phantom II (supersonic)
0.048	F-104 Starfighter
0.095	X-15 (Not confirmed)

Blunt and Streamlined Body Flows

Concept

Drag, in the context of fluid dynamics, refers to forces that act on a solid object in the direction of the relative flow velocity (note that the diagram below shows the drag in the opposite direction to the flow). The aerodynamic forces on a body come primarily from differences in pressure and viscous shearing stresses. Thereby, the drag force on a body could be divided into two components, namely frictional drag (viscous drag) and pressure drag (form drag). The net drag force could be decomposed as follows:

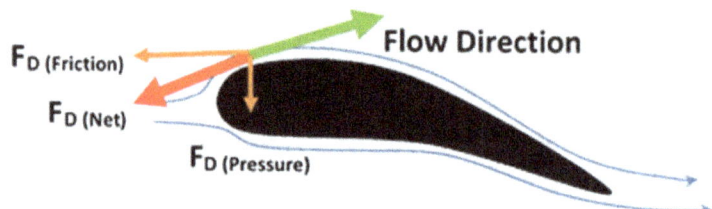

Flow across an airfoil showing the relative impact of drag force to the direction of motion of fluid over the body. This drag force gets divided into frictional drag and pressure drag. The same airfoil is considered as a streamlined body if friction drag (viscous drag) dominates pressure drag and is considered a blunt body when pressure drag (form drag) dominates friction drag.

$$c_{\mathrm{d}} = \frac{2F_{\mathrm{d}}}{\rho v^2 A} = c_{\mathrm{p}} + c_{\mathrm{f}} = \underbrace{\frac{1}{\rho v^2 A} \int_S \mathrm{d}A (p - p_o)\left(\hat{\mathbf{n}} \cdot \hat{\mathbf{i}}\right)}_{c_{\mathrm{p}}} + \underbrace{\frac{1}{\rho v^2 A} \int_S \mathrm{d}A \left(\hat{\mathbf{t}} \cdot \hat{\mathbf{i}}\right) T_w}_{c_{\mathrm{f}}}$$

where:

c_{p} is the pressure drag coefficient,

c_{f} is the friction drag coefficient,

$\hat{t}$ = Tangential direction to the surface with area dA,

$\hat{n}$ = Normal direction to the surface with area dA,

T_{w} is the shear Stress acting on the surface dA,

P_{o} is the pressure far away from the surface dA,

P is pressure at surface dA,

$\hat{i}$ is the unit vector in direction normal to the surface dA, forming a unit vector $\mathrm{d}\hat{A}$

Therefore, when the drag is dominated by a frictional component, the body is called a streamlined body; whereas in the case of dominant pressure drag, the body is called a blunt body. Thus, the shape of the body and the angle of attack determine the type of drag. For example, an airfoil is considered as a body with a small angle of attack by the fluid flowing across it. This means that it has attached boundary layers, which produce much less pressure drag.

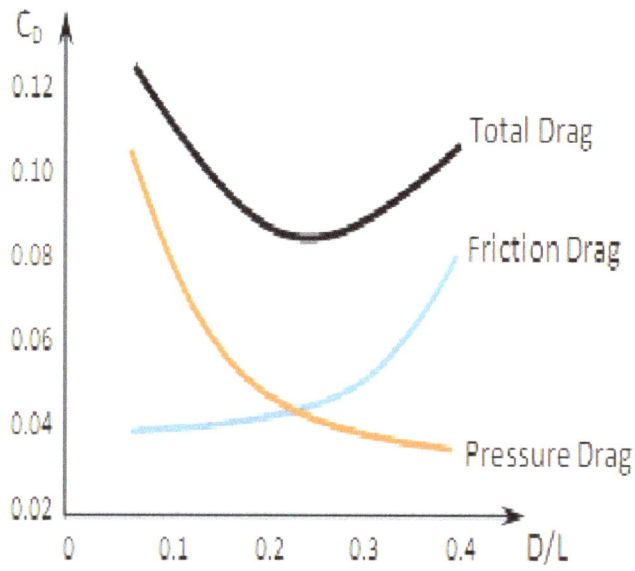

Trade-off relationship between pressure drag and friction drag

The wake produced is very small and drag is dominated by the friction component. Therefore, such a body (here an airfoil) is described as streamlined, whereas for bodies with fluid flow at high angles of attack, boundary layer separation takes place. This mainly occurs due to adverse pressure gradients at the top and rear parts of an airfoil.

Due to this, wake formation takes place, which consequently leads to eddy formation and pressure loss due to pressure drag. In such situations, the airfoil is stalled and has higher pressure drag than friction drag. In this case, the body is described as a blunt body.

A streamlined body looks like a fish (Tuna, Oropesa, etc.) or an airfoil with small angle of attack, whereas a blunt body looks like a brick, a cylinder or an airfoil with high angle of attack. For a given frontal area and velocity, a streamlined body will have lower resistance than a blunt body. Cylinders and spheres are taken as blunt bodies because the drag is dominated by the pressure component in the wake region at high Reynolds number.

To reduce this drag, either the flow separation could be reduced or the surface area in contact with the fluid could be reduced (to reduce friction drag). This reduction is necessary in devices like cars, bicycle, etc. to avoid vibration and noise production.

Practical Example

The aerodynamic design of cars has evolved from the 1920s to the end of the 20th century. This change in design from a blunt body to a more streamlined body reduced the drag coefficient from about 0.95 to 0.30.

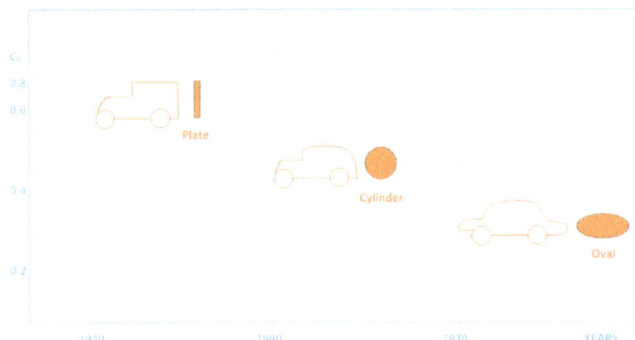

Time history of cars' aerodynamic drag in comparison to change in geometry
of streamlined bodies (blunt to streamline).

Drag Coefficient

The atmosphere and ocean exchange momentum, heat and mass including aerosols/tracers, which have a bearing on biological productivity of the oceans. The dynamical interaction between the ocean surface and wind is incorporated through roughness length, z_0, which depends on the wind at standard height (10 m) above the surface, wave formation and breaking and momentum dissipation at the surface. In the absence of waves all the momentum will be transferred to the sea surface. However, for momentum transfer, viscosity of a fluid defines its diffusive capacity. Therefore at the ocean-atmosphere interface, the ratio of the diffusive capacities of air and water directly gives the surface velocity in terms of the geostrophic wind U_g. For given density and kinematic viscosity ρ_a, μ_a of air and ρ_w, μ_w of seawater, the surface drift velocity U_0 on the ocean side of the interface is given by

$$U_0 = \frac{\rho_a}{\rho_w}\left(\frac{\nu_a}{\nu_w}\right)^{0.5} \times U_g = \frac{1}{200}U_g = 0.005U_g \qquad ....(1)$$

U_g has the same direction as that of geostrophic velocity U_g. However, the Ekman theory predicts the interface velocity in the direction of the resultant of geostrophic wind and the current vectors. This amounts to a contradiction between the two views and further studies are necessary. Below the interface into the sea, the mean velocity is much smaller than the interface drift velocity. However, for air-sea interaction, it is the 10-m wind (U_{10}), which is used in the calculations. Moreover, because surface roughness of the ocean waters changes with wind speed, a *drag coefficient* C_D may therefore be defined to deal with the changing winds and wave interaction. For wind speeds exceeding 8 ms^{-1} we have

$$C_D = \frac{u_*^2}{(U_{10} - U_0)^2}; \quad U = \sqrt{u_x^2 + u_y^2} \qquad ....(2)$$

where u_* is the *friction velocity*, U_{10} and U_0 are respectively the wind speeds at 10-m height and at the earth surface. The drag coefficient is an appropriate parameter for the momentum transfer under different atmospheric conditions and the state of the sea surface – stable, unsteady or dominated by waves. Once C_D is known, then from the boundary layer theory, the wind stress (τ) at the surface is defined as,

$$\tau = \rho_a u_*^2 = \rho_a C_D (U_{10} - U_0)^2 = \rho_a C_D U_{10}^2 \ as \ U_0 \ll U_{10} \ ....(3)$$

That is, the wind stress τ is directly related to the 10-m windspeed from (2). The formula (2) is one of the *bulk aerodynamic formulae* for computing the fluxes of momentum, heat and moisture at the atmosphere-ocean and atmosphere-land interface.

On the ground $U_0 = 0$, but over the ocean U_0 is the velocity of the wind at the sea surface. Though ocean is not a solid surface, yet the surface velocity U_0 is sufficiently small (typically $U_0 \approx 0.03 \ U_{10}$). This is basically due to the density differences between the air and water and the same momentum of air masses at air-sea interface can be carried with much smaller velocities of water masses. Hence the turbulence production over ocean is similar to land because the shear over the oceans is as large as that over the land. The drag coefficient C_D increases with wind speed (U) for the ocean surface as well, thus

$$10^3 C_D = \begin{cases} 1.1 & U_{10} \le 6.0 ms^{-1} \\ 0.61 + 0.0063 U_{10,} & 6.0 < U_{10} < 22.0 ms^{-1} \end{cases} \qquad (Smith \ 1980)$$

Charnock (1981) gave the following expression for the drag coefficient:

$$C_D = K \left\{ \ln \left[\frac{\rho g Z}{a \tau} \right] \right\}^{-1} ; k = 0.4, \ a = 0.0185$$

The parameter κ is the von Karman Constant. $U_{10} = \sqrt{u_{10}^2 + v_{10}^2}$ is the windspeed at 10 m height and U_0 is the component of ocean surface velocity along the wind direction. U_0 is generally neglected in comparison to U_{10}; but it should not be neglected in the regions of ocean where surface currents are strong and winds are weak.

Bulk Parameterization of Heat and Moisture

The expression (3) relates the wind stress τ with windspeed at 10 m height, which can be further simplified by neglecting U_0, which gives the following expression for the friction velocity

$$u_*^2 = C_D U_{10}^2 \ ....(4)$$

The bulk aerodynamic formulae for the sensible heat flux Q_H and the moisture flux E are given by

$$Q_H = -\rho_a C_p C_H U_{10} (T - T_0) \ ....(5)$$

$$E = -\rho_a C_E U_{10} [q(T) - q(T_0)] \ ....(6)$$

The latent heat flux is then calculated as

$$Q_E = L_v E \ ....(7)$$

These parameterizations are applicable both for land and sea with the coefficients C_H for sensible heat and C_E for water vapour. These parameters must be known for stable, neutral and unstable atmospheric conditions. Both C_H (Stanton number) and C_E (Dalton number) are the dimensionless aerodynamic exchange coefficients for temperature and humidity transfer respectively. When one uses these formulae for air-sea exchange to compute the fluxes of sensible heat and evaporation, then T and q in (8.5) and (8.6) are respectively replaced by the sea surface temperature T_S and saturated specific humidity q_S. Since the primary goal is to estimate the fluxes of momentum, heat and moisture using mean observations at one height, the formulae (8.4) – (8.6) fulfil this requirement.

The value of both C_H and C_E in neutral conditions is 1.2 x 10^{-3} which remains independent of wind-speeds within 5 - 20 ms^{-1}, for wind speeds exceeding 20 ms^{-1} possibly C_E may increase due to en-hanced evaporation. Under unstable conditions, the transfer coefficients have large values for light wind conditions, which decrease with increasing windspeeds. For C_H and C_E, Smith (1980) found a good fit of data with the following values

$$C_H = \begin{cases} 0.83 \times 10^{-3} & for \ \ stable \ \ conditions \\ 1.10 \times 10^{-3} & for \ \ unstable \ \ conditions \end{cases}$$

$$C_E = 1.50 \times 10^{-3}$$

The bulk aerodynamic formulae are based on the premise that the near-surface turbulence arises from mean wind shear over the surface and that turbulent fluxes of heat and moisture are propor-tional to their gradients just above the sea surface. Like the shear, the temperature and humidity gradients also increase, as the ocean surface is approached, in inverse proportion (~ 1/z) with the distance (z) from the surface. Both heat and moisture are vertically transferred at the air-sea inter-face by bodily movement of air parcels. Hot and moist air parcels move upward and relatively cold and dry air would be transferred downward.

Ekman Layer

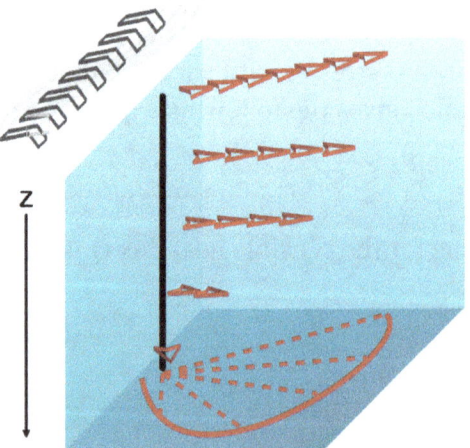

The Ekman layer is the layer in a fluid where the flow is the result of a balance between pressure gradient, Coriolis and turbulent drag forces. In the picture above, the wind blowing North creates a surface stress and a resulting Ekman spiral is found below it in the column of water.

The Ekman layer is the layer in a fluid where there is a force balance between pressure gradient force, Coriolis force and turbulent drag. It was first described by Vagn Walfrid Ekman. Ekman layers occur both in the atmosphere and in the ocean.

There are two types of Ekman layers. The first type occurs at the surface of the ocean and is forced by surface winds, which act as a drag on the surface of the ocean. The second type occurs at the bottom of the atmosphere and ocean, where frictional forces are associated with flow over rough surfaces.

History

Ekman developed the theory of the Ekman layer after Fridtjof Nansen observed that ice drifts at an angle of 20°-40° to the right of the prevailing wind direction while on an Arctic expedition aboard the Fram. Nansen asked his colleague, Vilhelm Bjerknes to set one of his students upon study of the problem. Bjerknes tapped Ekman, who presented his results in 1902 as his doctoral thesis.

Mathematical Formulation

The mathematical formulation of the Ekman layer begins by assuming a neutrally stratified fluid, a balance between the forces of pressure gradient, Coriolis and turbulent drag.

$$-fv = -\frac{1}{\rho_o}\frac{\partial p}{\partial x} + K_m\frac{\partial^2 u}{\partial z^2}, \qquad fu = -\frac{1}{\rho_o}\frac{\partial p}{\partial y} + K_m\frac{\partial^2 v}{\partial z^2}, \qquad 0 = -\frac{1}{\rho_o}\frac{\partial p}{\partial z},$$

where u and v are the velocities in the x and y directions, respectively, f is the local Coriolis parameter, and K_m is the diffusive eddy viscosity, which can be derived using mixing length theory. Note that p is a modified pressure: we have incorporated the hydrostatic of the pressure, to take account of gravity.

There are many regions where an Ekman layer is theoretically plausible; they include the bottom of the atmosphere, near the surface of the earth and ocean, the bottom of the ocean, near the sea floor and at the top of the ocean, near the air-water interface. Different boundary conditions are appropriate for each of these different situations. Each of these situations can be accounted for through the boundary conditions applied to the resulting system of ordinary differential equations. The separate cases of top and bottom boundary layers are shown below.

Ekman Layer at the Ocean (or free) Surface

We will consider boundary conditions of the Ekman layer in the upper ocean:

$$\text{at } z = 0: \quad A\frac{\partial u}{\partial z} = \tau^x \quad \text{and} \quad A\frac{\partial v}{\partial z} = \tau^y,$$

where τ^x and τ^y are the components of the surface stress, τ, of the wind field or ice layer at the top of the ocean.

For the boundary condition on the other side, as $z \to \infty : u \to u_g, v \to v_g$, where u_g and v_g are the geostrophic flows in the x and y directions.

Solution

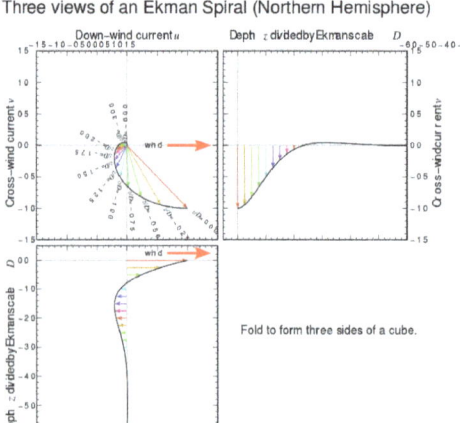

Three views of the wind-driven Ekman layer at the surface of the ocean in the Northern Hemisphere.
The geostrophic velocity is zero in this example.

These differential equations can be solved to find:

$$u = u_g + \frac{\sqrt{2}}{fd} e^{z/d} \left[\tau^x \cos(z/d - \pi/4) - \tau^y \sin(z/d - \pi/4) \right],$$

$$v = v_g + \frac{\sqrt{2}}{fd} e^{z/d} \left[\tau^x \sin(z/d - \pi/4) + \tau^y \cos(z/d - \pi/4) \right],$$

$$d = \sqrt{2K_m / |f|}$$

The value d is called the Ekman layer depth, and gives an indication of the penetration depth of wind-induced turbulent mixing in the ocean. Note that it varies on two parameters: the turbulent diffusivity K_m, and the latitude, as encapsulated by f. For a typical $K_m = 0.1 m^2/s$, and at 45° latitude ($f = 10^{-4}s^{-1}$), then d is approximately 45 meters. This Ekman depth prediction does not always agree precisely with observations.

This variation of horizontal velocity with depth ($-z$) is referred to as the Ekman spiral, diagrammed above and at right.

By applying the continuity equation we can have the vertical velocity as following

$$w = \frac{1}{f\rho_o} \left[-\left(\frac{\partial \tau^x}{\partial x} + \frac{\partial \tau^y}{\partial y} \right) e^{z/d} \sin(z/d) + \left(\frac{\partial \tau^y}{\partial x} - \frac{\partial \tau^x}{\partial y} \right) (1 - e^{z/d} \cos(z/d)) \right].$$

Note that when vertically-integrated, the volume transport associated with the Ekman spiral is to the right of the wind direction in the Northern Hemisphere.

Ekman Layer at the Bottom of the Ocean and Atmosphere

The traditional development of Ekman layers bounded below by a surface utilizes two boundary conditions:

- A no-slip condition at the surface;

- The Ekman velocities approaching the geostrophic velocities as goes to infinity.

Experimental Observations of the Ekman Layer

There is much difficulty associated with observing the Ekman layer for two main reasons: the theory is too simplistic as it assumes a constant eddy viscosity, which Ekman himself anticipated, saying and because it is difficult to design instruments with great enough sensitivity to observe the velocity profile in the ocean.

Laboratory Demonstrations

The bottom Ekman layer can readily be observed in a rotating cylindrical tank of water by dropping in dye and changing the rotation rate slightly. Surface Ekman layers can also be observed in rotating tanks.

In the Atmosphere

In the atmosphere, the Ekman solution generally overstates the magnitude of the horizontal wind field because it does not account for the velocity shear in the surface layer. Splitting the boundary layer into the surface layer and the Ekman layer generally yields more accurate results.

In the Ocean

The Ekman layer, with its distinguishing feature the Ekman spiral, is rarely observed in the ocean. The Ekman layer near the surface of the ocean extends only about 10 – 20 meters deep, and instrumentation sensitive enough to observe a velocity profile in such a shallow depth has only been available since around 1980. Also, wind waves modify the flow near the surface, and make observations close to the surface rather difficult.

Instrumentation

Observations of the Ekman layer have only been possible since the development of robust surface moorings and sensitive current meters. Ekman himself developed a current meter to observe the spiral that bears his name, but was not successful. The Vector Measuring Current Meter and the Acoustic Doppler Current Profiler are both used to measure current.

Observations

The first documented observations of an Ekman-like spiral in the ocean were made in the Arctic Ocean from a drifting ice flow in 1958. More recent observations include (not an exhaustive list):

- The 1980 Mixed Layer Experiment

- Within the Sargasso Sea during the 1982 Long Term Upper Ocean Study

- Within the California Current during the 1993 Eastern Boundary Current experiment

- Within the Drake Passage region of the Southern Ocean

- In the eastern tropical Pacific, at 2°N, 140°W, using 5 current meters between 5 and 25 meters depth. This study noted that the geostrophic shear associated with tropical stability waves modified the Ekman spiral relative to what is expected with horizontally uniform density.

- North of the Kerguelan Plateau during the 2008 SOFINE experiment

Common to several of these observations spirals were found to be 'compressed', displaying larger estimates of eddy viscosity when considering the rate of rotation with depth than the eddy viscosity derived from considering the rate of decay of speed.

Mixed Layer

The oceanic or limnological mixed layer is a layer in which active turbulence has homogenized some range of depths. The surface mixed layer is a layer where this turbulence is generated by winds, surface heat fluxes, or processes such as evaporation or sea ice formation which result in an increase in salinity. The atmospheric mixed layer is a zone having nearly constant potential temperature and specific humidity with height. The depth of the atmospheric mixed layer is known as the mixing height. Turbulence typically plays a role in the formation of fluid mixed layers.

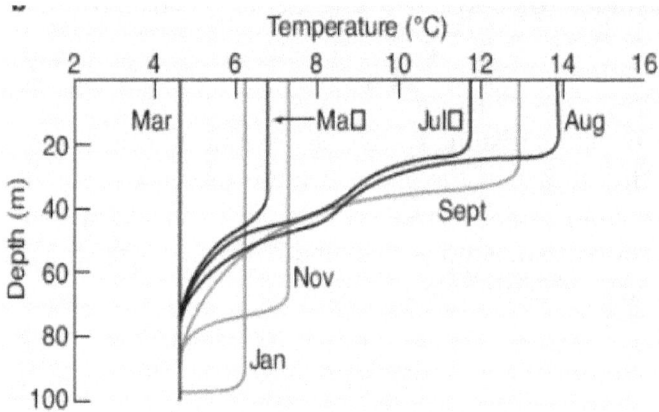

Depth of Mixed Layer versus temperature, along with relationship to different months of the year

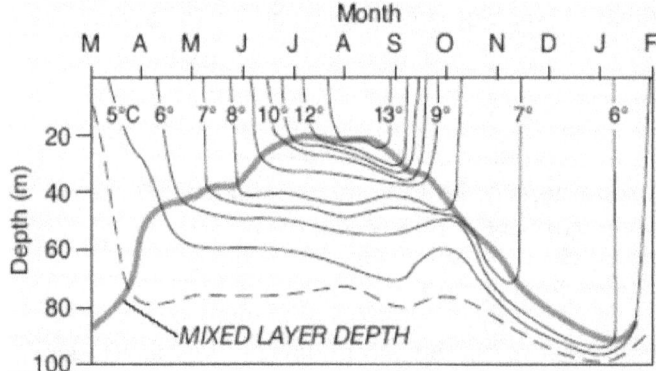

Depth of Mixed Layer versus the month of the year, along with relationship to temperature

Oceanic Mixed Layer

Importance of the Mixed Layer

The mixed layer plays an important role in the physical climate. Because the specific heat of ocean water is much larger than that of air, the top 2.5 m of the ocean holds as much heat as the entire atmosphere above it. Thus the heat required to change a mixed layer of 25 m by 1 °C would be sufficient to raise the temperature of the atmosphere by 10 °C. The depth of the mixed layer is thus very important for determining the temperature range in oceanic and coastal regions. In addition, the heat stored within the oceanic mixed layer provides a source for heat that drives global variability such as El Niño.

The mixed layer is also important as its depth determines the average level of light seen by marine organisms. In very deep mixed layers, the tiny marine plants known as phytoplankton are unable to get enough light to maintain their metabolism. The deepening of the mixed layer in the wintertime in the North Atlantic is therefore associated with a strong decrease in surface chlorophyll a. However, this deep mixing also replenishes near-surface nutrient stocks. Thus when the mixed layer becomes shallow in the spring, and light levels increase, there is often a concomitant increase of phytoplankton biomass, known as the "spring bloom".

Oceanic Mixed Layer Formation

There are three primary sources of energy for driving turbulent mixing within the open-ocean mixed layer. The first is the ocean waves, which act in two ways. The first is the generation of turbulence near the ocean surface, which acts to stir light water downwards. Although this process injects a great deal of energy into the upper few meters, most of it dissipates relatively rapidly. If ocean currents vary with depth, waves can interact with them to drive the process known as Langmuir circulation, large eddies that stir down to depths of tens of meters. The second is wind-driven currents, which create layers in which there are velocity shears. When these shears reach sufficient magnitude, they can eat into stratified fluid. This process is often described and modelled as an example of Kelvin-Helmholtz instability, though other processes may play a role as well. Finally, if cooling, addition of brine from freezing sea ice, or evaporation at the surface causes the surface density to increase, convection will occur. The deepest mixed layers (exceeding 2000 m in regions such as the Labrador Sea) are formed through this final process, which is a form of Rayleigh–Taylor instability. Early models of the mixed layer such as those of Mellor and Durbin included the final two processes. In coastal zones, large velocities due to tides may also play an important role in establishing the mixed layer.

The mixed layer is characterized by being nearly uniform in properties such as temperature and salinity throughout the layer. Velocities, however, may exhibit significant shears within the mixed layer. The bottom of the mixed layer is characterized by a gradient, where the water properties change. Oceanographers use various definitions of the number to use as the mixed layer depth at any given time, based on making measurements of physical properties of the water. Often, an abrupt temperature change called a thermocline occurs to mark the bottom of the mixed layer; sometimes there may be an abrupt salinity change called a halocline that occurs as well. The combined influence of temperature and salinity changes results in an abrupt density change, or pycnocline. Additionally, sharp gradients in nutrients (nutricline) and oxygen (oxycline) and a maximum in chlorophyll concentration are often co-located with the base of the seasonal mixed layer.

Oceanic Mixed Layer Depth Determination

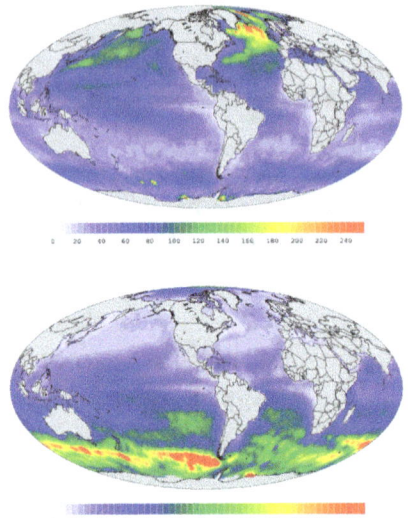

Mixed layer depth climatology for boreal winter(upper image) and boreal summer(lower image).

The depth of the mixed layer is often determined by hydrography—making measurements of water properties. Two criteria often used to determine the mixed layer depth are temperature and sigma-t (density) change from a reference value (usually the surface measurement). The temperature criterion used in Levitus (1982) defines the mixed layer as the depth at which the temperature change from the surface temperature is 0.5 °C. The sigma-t (density) criterion used in Levitus (1982) uses the depth at which a change from the surface sigma-t of 0.125 has occurred. Neither criterion implies that active mixing is occurring to the mixed layer depth at all times. Rather, the mixed layer depth estimated from hydrography is a measure of the depth to which mixing occurs over the course of a few weeks.

The mixed layer depth is in fact greater in winter than summer in each hemisphere. During the summer increased solar heating of the surface water leads to more stable density stratification, reducing the penetration of wind-driven mixing. Because seawater is most dense just before it freezes, wintertime cooling over the ocean always reduces stable stratification, allowing a deeper penetration of wind-driven turbulence but also generating turbulence that can penetrate to great depths.

Barrier Layer Thickness

The barrier layer thickness (BLT) is a layer of water separating the well-mixed surface layer from the thermocline. A more precise definition would be the difference between mixed layer depth (MLD) calculated from temperature minus the mixed layer depth calculated using density. The first reference to this difference as the barrier layer was in a paper describing observations in the western Pacific as part of the Western Equatorial Pacific Ocean Circulation Study. In regions where the barrier layer is present, stratification is stable because of strong buoyancy forcing associated with a fresh (i.e. more buoyant) water mass sitting on top of the water column.

In the past, a typical criterion for MLD was the depth at which the surface temperature cools by some change in temperature from surface values. For example, Levitus (1982) used 0.5°C. In the example to the right, 0.2°C is used to define the MLD (i.e. D_{T-02} in the Figure). Prior to the abundant subsurface salinity available from Argo, this was the main methodology for calculating the oceanic MLD. More recently, a density criterion has been used to define the MLD. The density-derived MLD is defined as the depth where the density increases from the surface value due to a prescribed temperature decrease of some value (e.g. 0.2°C) from the surface value while maintaining constant surface salinity value. In the Figure, this is defined by D_{sigma} and corresponds to a layer that is both isothermal and isohaline. The BLT is the difference of the temperature-defined MLD minus the density-defined value (i.e. D_{T-02} - D_{sigma}).

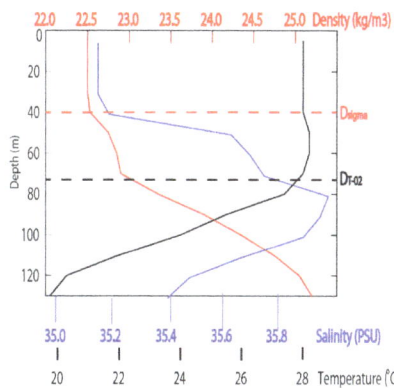

An example of barrier layer thickness for an Argo profile taken January 31, 2002 in the tropical Indian Ocean. The red line is the density profile, black line is temperature, and the blue line is salinity. One mixed layer depth, D_{T-02}, is defined as the depth at which the surface temperature cools by 0.2°C (black dashed line). The density defined mixed layer, D_{sigma}, is 40 m (red dashed line) and is defined as the surface density plus the density difference brought about by the temperature increment of 0.2°C. Above D_{sigma} the water is both isothermal and isohaline. The difference between D_{T-02} minus D_{sigma} is the barrier layer thickness (blue arrows on the figure) .

BLT Regimes

Large values of the BLT are typically found in the equatorial regions and can be as high as 50 m. Above the barrier layer, the well mixed layer may be due to local precipitation exceeding evaporation (e.g. in the western Pacific), monsoon related river runoff (e.g. in the northern Indian Ocean), or advection of salty water subducted in the subtropics (found in all subtropical ocean gyres). Barrier layer formation in the subtropics is associated with seasonal change in the mixed layer depth, a sharper gradient in sea surface salinity (SSS) than normal, and subduction across this SSS front. In particular, the barrier layer is formed in winter season in the equatorward flank of subtropical salinity maxima. During early winter, the atmosphere cools the surface and strong wind and negative buoyancy forcing mixes temperature to a deep layer. At this same time, fresh surface salinity is advected from the rainy regions in the tropics. The deep temperature layer along with strong stratification in the salinity gives the conditions for barrier layer formation.

For the western Pacific, the mechanism for barrier layer formation is different. Along the equator, the eastern edge of the warm pool (typically 28°C isotherm - see SST plot in the western Pacific) is a demarcation region between warm fresh water to the west and cold, salty, upwelled water in the central Pacific. A barrier layer is formed in the isothermal layer when salty water is subducted (i.e. a denser water mass moves below another) from the east into the warm pool due to local convergence or warm fresh water overrides denser water to the east. Here, weak winds, heavy precipitation, eastward advection of low salinity water, westward subduction of salty water and downwelling equatorial Kelvin or Rossby waves are factors that contribute to deep BLT formation.

Importance of BLT

Prior to El Nino, the warm pool stores heat and is confined to the far western Pacific. During the El Nino, the warm pool migrates eastward along with the concomitant precipitation and current anomalies. The fetch of the westerlies is increased during this time, reinforcing the event. Using data from the ship of opportunity and Tropical Atmosphere – Ocean (TAO) moorings in the western Pacific, the east and west migration of the warm pool was tracked over 1992-2000 using sea surface salinity (SSS), sea surface temperature (SST), currents, and subsurface data from Conductivity, temperature, depth taken on various research cruises. This work showed that during westward flow, the BLT in the western Pacific along the equator (138°E-145°E, 2°N-2°S) was between 18 m – 35 m corresponding with warm SST and serving as an efficient storage mechanism for heat. Barrier layer formation is driven by westward (i.e. converging and subducting) currents along the equator near the eastern edge of the salinity front that defines the warm pool. These westward currents are driven by downwelling Rossby waves and represent either a westward advection of BLT or a preferential deepening of the deeper thermocline versus the shallower halocline due to Rossby wave dynamics (i.e. these waves favor vertical stretching of the upper water column). During El Nino, westerly winds drive the warm pool eastward allowing fresh water to ride on top of the local colder/saltier/denser water to the east. Using coupled, atmospheric/ocean models and tuning the mixing to eliminate BLT for one year prior to El Nino, it was shown that the heat buildup associated with barrier layer is a requirement for big El Nino. It has been shown that there is a tight relationship between SSS and SST in the western Pacific and the barrier layer is instrumental in maintaining heat and momentum in the warm pool within the salinity stratified layer. Later work, including Argo drifters, confirm the relationship between eastward migration of the warm pool during El Nino and barrier layer heat storage in the western Pacific. The main impact of barrier layer is to maintain a shallow mixed layer allowing an enhanced air-sea coupled response. In addition, BLT is the key factor in establishing the mean state that is perturbed during El Nino/La Nina

Limnological Mixed Layer Formation

Formation of a mixed layer in a lake is similar to that in the ocean, but mixing is more likely to occur in lakes solely due to the molecular properties of water. Water changes density as it changes temperature. In lakes, temperature structure is complicated by the fact that fresh water is heaviest at 3.98 °C (degrees Celsius). Thus in lakes where the surface gets very cold, the mixed layer briefly extends all the way to the bottom in the spring, as surface warms as well as in the fall, as the surface cools. This overturning is often important for maintaining the oxygenation of very deep lakes.

The study of limnology encompasses all inland water bodies, including bodies of water with salt in them. In saline lakes and seas (such as the Caspian Sea), mixed layer formation generally behaves similarly to the ocean.

Atmospheric Mixed Layer Formation

The atmospheric mixed layer results from convective air motions, typically seen towards the middle of the day when air at the surface is warmed and rises. It is thus mixed by Rayleigh–Taylor instability. The standard procedure for determining the mixed layer depth is to examine the profile of potential temperature, the temperature which the air would have if it were brought to the pressure found at the surface without gaining or losing heat. As such an increase of pressure involves compressing the air, the potential temperature is higher than the in-situ temperature, with the difference increasing as one goes higher in the atmosphere. The atmospheric mixed layer is defined as a layer of (approximately) constant potential temperature, or a layer in which the temperature falls at a rate of approximately 10 °C/km, provided it is free of clouds. Such a layer may have gradients in the humidity, though. As is the case with the ocean mixed layer, velocities will not be constant throughout the atmospheric mixed layer.

References

- Chereskin, T.K. (1995). "Direct evidence for an Ekman balance in the California Current". Journal of Geophysical Research. 100: 18261–18269. Bibcode:1995JGR...10018261C. doi:10.1029/95JC02182

- J. S. Dugdale (1996). Entropy and its Physical Meaning. Taylor & Francis. p. 13. ISBN 0-7484-0569-0. This law is the basis of temperature

- Bosc, C.; Delcroix, T.; Maes, C. (2009). "Barrier layer variability in the western Pacific warm pool from 2000 to 2007". Journal of Geophysical Research: Oceans. 114. Bibcode:2009JGRC..114.6023B. doi:10.1029/2008jc005187

- Hawking, SW (1985). "Arrow of time in cosmology". Phys. Rev. D. 32 (10): 2489–2495. Bibcode:1985PhRvD..32.2489H. doi:10.1103/PhysRevD.32.2489. Retrieved 2013-02-15

- Lenn, Y; Chereskin, T.K. (2009). "Observation of Ekman Currents in the Southern Ocean". Journal of Physical Oceanography. 39: 768–779. Bibcode:2009JPO....39..768L. doi:10.1175/2008jpo3943.1

- Sychev, V. V. (1991). The Differential Equations of Thermodynamics. Taylor & Francis. ISBN 978-1-56032-121-7. Retrieved 2012-11-26

- Lebowitz, Joel L. (September 1993). "Boltzmann's Entropy and Time's Arrow" (PDF). Physics Today. 46 (9): 32–38. Bibcode:1993PhT....46i..32L. doi:10.1063/1.881363. Retrieved 2013-02-22

- Kato, H.; Phillips, O.M. (1969). "On the penetration of a turbulent layer into a stratified fluid". J. Fluid Mechanics. 37: 643–655. doi:10.1017/S0022112069000784

- Cronin, M.F.; Kessler, W.S. (2009). "Near-Surface Shear Flow in the Tropical Pacific Cold Tongue Front". Journal of Physical Oceanography. 39: 1200–1215. doi:10.1175/2008JPO4064.1

- Truesdell, C., Muncaster, R.G. (1980). Fundamentals of Maxwell's Kinetic Theory of a Simple Monatomic Gas, Treated as a Branch of Rational Mechanics, Academic Press, New York, ISBN 0-12-701350-4, p. 15

- Mellor, G. L.; Durbin, P. A. (1975). "The structure and dynamics of the ocean surface mixed layer". Journal of Physical Oceanography. 5: 718–728. doi:10.1175/1520-0485(1975)005<0718:TSADOT>2.0.CO;2

- van Gool, W.; Bruggink, J.J.C. (Eds) (1985). Energy and time in the economic and physical sciences. North-Holland. pp. 41–56. ISBN 0-444-87748-7

- Craik, A.D.D.; Leibovich, S. (1976), "A Rational model for Langmuir circulations", Journal of Fluid Mechanics, 73: 401–426, Bibcode:1976JFM....73..401C, doi:10.1017/S0022112076001420

- Grandy, W.T., Jr (2008). Entropy and the Time Evolution of Macroscopic Systems, Oxford University Press, Oxford, ISBN 978-0-19-954617-6, pp. 55–58

- Agrawal, Y.C.; Terray, E.A.; Donelan, M.A.; Hwang, P.A.; Williams, A.J.; Drennan, W.M.; Kahma, K.K.; Kitaiigorodski, S.A. "Enhanced dissipation of kinetic energy beneath surface waves". Nature. 359: 219–220. doi:10.1038/359219a0

- Sprintall, J., and M. Tomczak, Evidence of the barrier layer in the surface-layer of the tropics, Journal of Geophysical Research: Oceans, 97 (C5), 7305-7316, 1992

- Cushman-Roisin, Benoit (1994). "Chapter 5 - The Ekman Layer". Introduction to Geophysical Fluid Dynamics (1st ed.). Prentice Hall. pp. 76–77. ISBN 0-13-353301-8

- Maes, C.; Belamari, S. (2011). "On the Impact of Salinity Barrier Layer on the Pacific Ocean Mean State and ENSO". Sola. 7: 97–100. doi:10.2151/sola.2011-025

- Lukas, R.; Lindstrom, E. (1991). "The Mixed Layer of the Western Equatorial Pacific-Ocean". Journal of Geophysical Research: Oceans. 96: 3343–3357. Bibcode:1991JGR....96.3343L. doi:10.1029/90jc01951

- Vallis, Geoffrey K. (2006). "Chapter 2 – Effects of Rotation and Stratification". Atmospheric and Oceanic Fluid Dynamics (1st ed.). Cambridge, UK: Cambridge University Press. pp. 112–113. ISBN 0-521-84969-1

- Weller, R.A.; Davis, R.E. (1980). "A vector-measuring current meter". Deep-Sea Research. 27 (7): 565–582. Bibcode:1980DSRI...27..565W. doi:10.1016/0198-0149(80)90041-2

- Mignot, J., C.d.B. Montegut, A. Lazar, and S. Cravatte, Control of salinity on the mixed layer depth in the world ocean: 2. Tropical areas, Journal of Geophysical Research: Oceans, 112 (C10), 2007

- Holton, James R. (2004). "Chapter 5 – The Planetary Boundary Layer". Dynamic Meteorology. International Geophysics Series. 88 (4th ed.). Burlington, MA: Elsevier Academic Press. pp. 129–130. ISBN 0-12-354015-1

- Hunkins, K. (1966). "Ekman drift currents in the Arctic Ocean". Deep-Sea Research. 13: 607–620. Bibcode:1966DSROA..13..607H. doi:10.1016/0011-7471(66)90592-4

- Delcroix, T.; McPhaden, M. (2002). "Interannual sea surface salinity and temperature changes in the western Pacific warm pool during 1992-2000". Journal of Geophysical Research: Oceans. 107 (C12)

- Maes, C.; Picaut, J.; Belamari, S. (2005). "Importance of the salinity barrier layer for the buildup of El Niño". Journal of Climate. 18 (1): 104–118. doi:10.1175/jcli-3214.1

Permissions

Index